T0323478

Industrial Automation and Robotics

This book discusses the radical technological changes occurring due to Industry 4.0, with a focus on offering a better understanding of the Fourth Industrial Revolution. It also presents a detailed analysis of interdisciplinary knowledge, numerical modeling and simulation, and the application of cyber–physical systems, where information technology and physical devices create synergic systems leading to unprecedented efficiency.

The book focuses on industrial applications of automation and robotics. It covers recent developments and trends occurring in both computer-aided manufacturing techniques, as well as computer-aided assembly techniques. Robots using embedded systems and artificial intelligence applications are also covered.

Industrial Automation and Robotics: Techniques and Applications offers theoretical results, practical solutions, and guidelines that are valuable for both researchers and those working in the area of engineering.

Industrial Automation and Robotics
Techniques and Applications

Edited by
Kaushik Kumar and Sridhar B. Babu

CRC Press
Taylor & Francis Group
Boca Raton   London   New York

CRC Press is an imprint of the
Taylor & Francis Group, an **informa** business

MATLAB® is a trademark of The MathWorks, Inc. and is used with permission. The MathWorks does not warrant the accuracy of the text or exercises in this book. This book's use or discussion of MATLAB® software or related products does not constitute endorsement or sponsorship by The MathWorks of a particular pedagogical approach or particular use of the MATLAB® software.

First edition published 2023
by CRC Press
6000 Broken Sound Parkway NW, Suite 300, Boca Raton, FL 33487-2742

and by CRC Press
4 Park Square, Milton Park, Abingdon, Oxon, OX14 4RN

CRC Press is an imprint of Taylor & Francis Group, LLC

ISBN: 978-0-367-48797-3 (hbk)
ISBN: 978-0-367-63985-3 (pbk)
ISBN: 978-1-003-12164-0 (ebk)

DOI: 10.1201/9781003121640

Typeset in Times LT Std
by KnowledgeWorks Global Ltd.

Contents

Preface

The editors are pleased to present the book *Industrial Automation and Robotics: Techniques and Applications*. The book title was chosen understanding the current importance of the subject.

Automation and robotics are a specialized engineering stream that provides in-depth knowledge in areas concerning electro-mechanics, robotic sensors, automatic system, and artificial intelligence. The aim of automation is to boost efficiency and reliability. In most cases, however, automation replaces labor. In fact, economists today fear that new technology will eventually push up unemployment rates significantly. The term "manufacturing" refers to converting raw materials and components into finished goods, usually on a large scale in a factory and in many manufacturing plants today, robotic assembly lines are progressively carrying out functions that humans used to do. Automation encompasses many key elements, systems, and job functions in virtually all industries. It is especially prevalent in manufacturing, transportation, facility operations, and utilities. Additionally, national defense systems are becoming increasingly automated. Automation today exists in all functions within industry including integration, installation, procurement, maintenance, and even marketing and sales.

This book would be a key to support in elaborating the radical technological changes occurring due to the Industry 4.0, with a focus on offering a better understanding of the Fourth Industrial Revolution. It also presents a detailed analysis of interdisciplinary knowledge, numerical modeling, and simulation, as well as the application of cyber–physical systems, where information technology and physical devices create synergic systems leading to unprecedented efficiency. The theoretical results, practical solutions, and guidelines presented are valuable for both researchers working in the area of engineering sciences and practitioners looking for solutions to industrial problems.

As each chapter has been written by recognized experts and academicians practicing and following the outlook of the concept and, hence, the book contains many of "state-of-the-art" techniques required for practical applications. Thus, this book should serve as a useful source of information for practicing academicians and specialists, as well as for academic institutions working on the subject and last but not the least the industrial fraternity. At the core of the book are several application areas where concept of automation can be applied, each treated as a complete chapter. This book also provides an introduction to the approach and mindset that give the readers a flavor on the techniques and methods of the same. Application chapters provide discussion and reflection that will lead the reader into a deeper understanding of the nature of robotics with an industrial flavor. We are confident that this book will serve as an easy-to-understand guide to facilitate anyone to learn and apply the methods and tools to generate innovative ideas and allow to grasp the low-hanging fruits in the short term and design a better system in the long run for the betterment of industry in particular and society as a whole.

In order to understand the multidimensional aspect of the subject, the chapters of the book are segregated in *three sections*, namely *Section I: Industrial Engineering,*

Section II: Automation and Optimization, and *Section III: Robotics. Section I* contains *Chapters 1 and 2, Section II* comprises Chapters 3 and 4, and *Section III* consists of *Chapters 5–9*.

Section I of the book starts with *Chapter 1* providing extensive study on Internet of Things (IoT)-based embedded sensor system for real-time health monitoring of composite structures for large-scale industrial operations. The chapter provides proposal, definition, construction, and testing of the real-time monitoring system using a data acquisition and transmission system used to real-time health monitors the life cycle of composite structures using IoT-based embedded sensor systems, and it concludes that Industrial IoT (IIoT) technology is capable of detecting and monitoring composite structures, as well as interfacing with a personal-built cloud server profile via the IoT-based embedded sensor system.

Chapter 2 elaborately discusses a computer-aided ergonomic design and assembly of a domestic string hopper machine used exclusively by hoteliers and industrialists for making certain delicious dishes. The conventional machine can be replaced with the advanced machine for the extrusion of flour paste into strings. Existing machines are human-powered and are subjected to musculoskeletal disorders (MSDs). Therefore, the purpose of the work was to address the ergonomic risks imposed by this traditional practice and to design and develop an ergonomically pleasing electric string hopper machine for industrial and domestic purposes. The product, thus developed, was ergonomically pleasing and proved that it would prevent musculoskeletal disorders.

Chapter 3, the first chapter of *Section II*, provides optimal process parameters for performance enhancement in an automated nontraditional machining process. The chapter utilizes grey wolf optimizer (GWO) algorithm, a bio-inspired optimization technique to identify an optimal solution towards optimization of electrochemical machining (ECM) process parameters to improve machining quality. The chapter provides optimized ECM process parameters to meet the required degree of machining quality features using both single-objective and multi-objective optimization scenarios.

Chapter 4, the last chapter of the *section*, provides simulation for decision making in the orthogonal turning operation of three different aluminium alloys using John-Cook plasticity model and damage laws. The chapter achieves the set target of three-dimensional finite element simulation of unsteady-state metal cutting (orthogonal turning) of three different aluminium alloys AA-2024, AA-6061, and AA-7075. These simulations are performed in ABAQUS® FEA Software and for the three aluminium alloys under consideration the stresses, strains, cutting forces, and temperature distribution are compared and the easiest material for machining is reasoned out and derived based on the model. For validation of the work, comparisons with experimental investigations are also presented. The chapter predicts on material behavior during turning through the FE analyses and also highlights with special relevance to the chip formation, plasticization, and strain localization that result from adiabatic heating.

From here on the book talks about robotics. In *Chapter 5*, that is, the first chapter of the section dedicated to robotics develops forward kinematics of a two-DOF manipulator in MATLAB® environment. It also simultaneously provides an insight into MATLAB® application in the domain of robotics. As robotics is a multidisciplinary subject of study and involves intensive computations hence some external sophisticated computational tool that provides high efficiency and accuracy in computation is very much in demand. MATLAB® provides one such computational environment to serve the purpose. The chapter focuses on providing a strong foundation of robot kinematics and provides a lucid explanation of homogeneous transformations of a rigid body using the specific Denavit-Hartenberg links and joints parameters or notations. On the basis of DH notations, the homogeneous transformation matrix for each link of a two-DOF robot manipulator has been obtained using MATLAB® programming. It is worthy to note that the theoretical foundations provided in this chapter play an important role in a better understanding of the same thing implemented on the MATLAB® windows-based environment.

Chapter 6 highlights application of machine learning algorithms in intellectual robotics. A robot may resemble a human in activity and behavior or it may be a robotic process automation which simulates how humans engage with software to perform rule-based tasks. The learning experiences in robotics are sequenced to depict the growth of the learning process in which the machine is trained to learn new skills and during the learning process, machine learning tries to optimize memory, compute power, and speed. The task of finding a feasible and efficient training model for robotics is a major difficulty for machine learning process. The chapter proposes that the difficulty level can be reduced by incorporating an autonomous agent in the training stages which will learn about the external world independently over time and will create a set of complicated abilities and knowledge incrementally which in-turn makes a successful machine learning model.

One of the primary aspects of Industry 4.0 is 3D printing (3DP), *Chapter 7* provides an amalgamated version of 3D printing and robotics as applied to healthcare industry. 3DP of biological tissues is currently under priority research and would lead to the fabrication of human organs in the future and robotics are projected to play an increasing role in personalized medicine in order to make it a reality. Hence, production of items like pre-filled needles, robot technology is currently being used. The book chapter exclusively focuses on the role of 3DP, emerging technologies in pharmaceutical industries, latest applications in healthcare, regulatory concerns, robotics in healthcare industry, and many more aspects of interdisciplinary/multidisciplinary research and innovation reports in the field of 3DP, machine learning, artificial intelligence, coupled with robotics.

Chapter 8 provides an experimental learning of self-repeating robotic arm. The efficacy of the work can be attributed to the fact that repeated actions require more time and energy and also becomes monotonous. But there are several situations in which a repeated instruction or an action is required. For instance, in the wake of current pandemic of COVID-19, it was mandatory to issue safety instructions, such as wearing a mask and social distancing. On an individual front, it is seamlessly an easy job; however, if this has to be given to a large number of people, then a robot or a robotic arm is definitely helpful to overcome many human constraints. In addition, many people are unable to move from one location to another due to their physiological or biological conditions, but they have to move in order to fetch various items. They need assistance to accomplish this; however, human assistance is not possible always. Thus, keeping the above two constraints in mind, it is decided to develop a robotic arm. The chapter depicts all the steps right from start to end for creating a robotic arm using an Arduino Uno R3, allowing the user to give five recurrent actions, which it records and then repeats until stopped.

The last chapter of the section and the book, *Chapter 9* describes robotic applications and its impact in the global agricultural sectors for advancing automation. Farmers in the global agriculture sector are benefiting from robotics applications. The agriculture sector, as one of the world's most important industries, is encouraging the hi-tech industry to develop new applications using major disruptive technologies such as robotics and artificial intelligence. Farmers can focus more on boosting total output yield by using agricultural robots to automate slow, repetitive, and boring jobs. Growing population and labor shortages, expanding IoT and navigation technology, and the COVID-19 pandemic are just a few of the causes driving the adoption of autonomous mobile robots in agriculture. The agriculture industry is now being impacted by issues such as changing demographics and urbanization. With the rapid development of machine vision and intelligent systems that are merged into mobile robots, the goal of designing self-directed systems accomplished of instable machinist happenings in global agricultural sectors is becoming a reality. The collaboration of autonomous mobile robots to complete one or more agricultural tasks is a step advance in the latest applications of autonomous mobile robots. This chapter outlines an overview of robotic applications and its impact for the advancement of automation in the wide agricultural industries.

First and foremost, we would like to thank God. It was His blessings that provided us the strength to believe in passion, hard work, and pursue our dreams. We thank our family for having the patience with us for taking yet another challenge which decreases the amount of time we spend with them. They were our inspiration and motivation. We would like to thank our parents and grandparents for allowing us to follow our ambitions. We would like to thank all of our colleagues, friends, known associates in different part of the world for sharing ideas in shaping our thoughts. Our efforts will come to a level of satisfaction if the students, researchers, and professionals concerned with all the fields related to industrial automation and robotics gets benefitted.

We, from the bottom of our heart, owe a huge thanks to each and every contributing authors, reviewers, editorial advisory board members, book development editor, and the team at CRC Press for their availability for work on this huge project. All of their efforts were instrumental in compiling this book and without their constant and consistent guidance, support, and cooperation, we wouldn't have reached this milestone. Especially, during this global pandemic, when all supports were withdrawn, we were elated to find the whole team of CRC Press by our side to support. We salute their dedication.

Last, but definitely not least, we would like to thank Ranjan Kumar, scholar of Birla Institute of Technology, Mesra, India, and all individuals who had taken the time out and helped us during the process of writing this book; without their support and encouragement, we would have probably given up the project.

About the Editors

Kaushik Kumar, BTech (Mechanical Engineering, REC [Now NIT], Warangal), MBA (Marketing, IGNOU) and PhD (Engineering, Jadavpur University), is presently an associate professor in the Department of Mechanical Engineering, Birla Institute of Technology, Mesra, Ranchi, India. He has 14 years of teaching and research and over 11 years of industrial experience in a manufacturing unit of global repute. His areas of teaching and research interest are quality management systems, optimization, non-conventional machining, CAD/CAM, rapid prototyping, and composites. He has 9 patents, 28 books, 17 edited book volumes, 43 book chapters, 137 international journals, 21 international, and 1 national conference publications to his credit. He is on the editorial board and review panel of seven international and one national journals of repute. He has been felicitated with many awards and honors.

B. Sridhar Babu has completed a BE with Mechanical Engineering from Kakatiya University, MTech with advanced manufacturing systems from JNTUH University and a PhD with Mechanical Engineering from JNTUH University. He has 21 years of teaching experience, 10 years of which were at CMR Institute of Technology itself. He joined the Department of Mechanical Engineering in CMR Institute of Technology on 1 July 2009. He is Fellow of the Institution of Engineers (I), Kolkata, and also member of ISTE, IAENG, and SAE India. He has published 38 papers in various international/national journals and international/national conferences. He is the author of four text books. He received Bharath Jyothi award for his research excellence from India international friendship society New Delhi, India. He is a paper setter for various universities, reviewer for various international journals and conferences, and guided more than 75 BTech and MTech projects. His research interests include manufacturing, advanced materials, mechanics of materials, and so on. He is a guest editor for proceedings of first International Conference on Manufacturing, Material Science and Engineering (ICMMSE 2019), Materials Today – Proceedings (Scopus and CPCI Indexed), and AIP Proceedings (Scopus Indexed).

Contributors

B. Acherjee
Birla Institute of Technology
Mesra, Ranchi, India

Bhargav Avinash G
Vellore Institute of Technology
Vellore, India

Ashrith Gadeela
Geethanjali College of Engineering and
    Technology
Cheeryal, Hyderabad, India

S. Ghosh
Jadavpur University
Kolkata, India

Nithya Jayakumar
K S Rangasamy College of Technology
Tiruchengode, India

Ramya Jayakumar
Hindusthan College of Engineering and
    Technology
Coimbatore, India

Padmanabhan K
Vellore Institute of Technology
Vellore, India

Atul Kadam
Shree Santakrupa College of
    Pharmacy
Ghogaon, India

Prachi Khamkar
Next Big Innovation Labs
Bangalore, India

A.S. Kuar
Jadavpur University
Kolkata, India

A. Kishore Kumar
Sri Ramakrishna Engineering College
Coimbatore, India

Kaushik Kumar
Birla Institute of Technology
Mesra, Ranchi, India

Ranjan Kumar
Birla Institute of Technology
Mesra, Ranchi, India

Raghu Raja Pandiyan Kuppusamy
National Institute of Technology
Warangal, India

Prudvi Krishna M
Vellore Institute of Technology
Vellore, India

Saraschandra M
Vellore Institute of Technology
Vellore, India

Debarshi Kar Mahapatra
Dadasaheb Balpande College of
    Pharmacy
Nagpur, India

G. Madhan Mohan
PSG College of Technology
Coimbatore, India

B. Nagamani
Geethanjali College of Engineering and
    Technology
Cheeryal, Hyderabad, India

T. Nivethitha
Hindusthan College of Engineering and
    Technology
Coimbatore, India

Yashwanth Padarthi
Indian Institute of Technology
Kharagpur, India

Sathwik Parasara
Geethanjali College of Engineering and
Technology
Cheeryal, Hyderabad, India

P.K. Poonguzhali
Hindusthan College of Engineering and
Technology
Coimbatore, India

M. Vignesh Raja
PSG College of Technology
Coimbatore, India

Hari Sarada
Geethanjali College of Engineering and
Technology
Cheeryal, Hyderabad, India

N. Subadra
Geethanjali College of Engineering and
Technology
Cheeryal, Hyderabad, India

E. Selva Vignesh
PSG College of Technology
Coimbatore, India

C. Vigneswaran
PSG College of Technology
Coimbatore, India

Section I

Industrial Engineering

1 IoT-Based Embedded Sensor System for Real-Time Health Monitoring of Composite Structures for Large-Scale Industrial Operations

Yashwanth Padarthi
Indian Institute of Technology, Kharagpur,
West Bengal, India

Raghu Raja Pandiyan Kuppusamy
National Institute of Technology, Warangal,
Telangana, India

CONTENTS

DOI: 10.1201/9781003121640-2

1.1   INTRODUCTION

Industrial Internet of Things (IIoT) technology is a combination of several technologies that have been in the industrial backdrop for many years, such as machine
learning, big data, sensor data for communication, and automation [1, 2]. The IIoT
encompasses a wide range of IoT applications in industry, such as smart product
design concepts and data-driven automation methods in the industrial sector. It
extensively employs contemporary sensor technology to improve many types of
equipment with remote monitoring and maintenance capabilities [3]. The IIoT is a
critical component of the smart factory, an automation trend that combines current
cloud computing, IIoT, and Artificial Intelligence (AI) [4]. IIoT is a critical component of a smart factory, an automation trend that combines current cloud computing,
IIoT, and AI to build intelligent, self-optimizing industrial equipment and facilities
[5]. Separate sections of a production line that are enabled with IoT interact with
each other in near real time, making the whole manufacturing process much easier
to monitor and operate. Finally, IIoT also encourages more adaptable, open architectures that allow for more customization and digital updates across tens of thousands
of devices [6]. The IIoT can protect industrial systems from downtime by identifying
early indicators of trouble and preventing them from worsening [6]. Installing IoT
sensors in industrial systems enables operators to understand what is going on with
the equipment in near real time and with great precision.

The IoT is a novel concept that uses wireless/wired technology to link the ubiquitous existence of a number of things or artefacts around us to the Internet in order
to achieve monitoring of the composite structures instrumented with distributed
optical fiber sensors (DOFS) [7]. Since the idea of IoT was launched in 2005, variety of applications such as the deployment of a new generation of networked smart
objects with connectivity along with various visual and action capabilities have been
enabled [8]. The IoT technology is of utmost importance in various applications
involving monitoring of hostile environments [9]. This chapter describes a revamped
task-technology fit approach to examine how IoT technology can be integrated and
optimize emergency response operations, utilizing the data necessary for the identification of the information requirements in terms of well-organized assistance, precise situational alertness, and wide-ranging visibility of resources monitored [10].
Monitoring the health of composite structures can benefit and enhance the structural
monitoring applications as shown in Figure 1.1. However, damage to the composite
structures does not indicate a complete failure of functionality; rather, it indicates

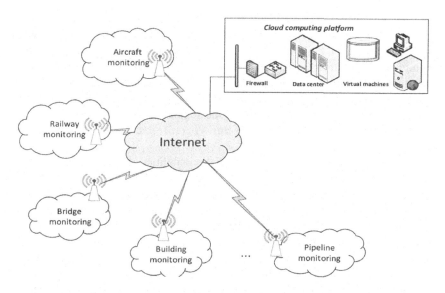

FIGURE 1.1   Schematic of Internet of Things (IoT)-based health monitoring system.

that the composite system is no longer in optimal condition, and the structure will collapse if the damage increases.

Health monitoring is a method of detecting damage gathering data from appropriate sensors on a regular basis to characterize the damage and determine the structure's health status. As a result, this monitoring will provide information on the structure's status, especially its integrity, in a timely manner, and it is important to evaluate performance and health status for composite structures instrumented with DOFS. The aim of this project is to build an IoT framework for health monitoring in order to identify potential damages and track how the composite structure performs over time. A specialized server has been used for this purpose [11]. Via internet communication, the server may acquire data from the local host and synchronize acquisitions with DOFS. It also has a built-in security system and storage to permit post-processing of the acquired signal response from the composites with DOFS. The data acquired from the DOFS system has been utilized to characterize the damage in the composite structures instrumented with DOFS. In the present work, the experimental results show that the estimated damage increases when the structure is subjected to static and dynamic loading, impact loading, and corrosive environments.

1.2   LITERATURE REVIEW

Several works have demonstrated the significance of IoT-based sensor systems for real-time health monitoring of industrial operations in a variety of engineering areas. Morales et al. [12] monitored the temperature and humidity of a commercial-size greenhouse in Mexico for six months using an IoT-based health monitoring system. The authors also included a data-driven forecasting algorithm for greenhouse micro-climate conditions with the temperature forecasted 24 hours in advance accurately.

The results suggested that IoT framework will assist farmers in monitoring crops and enabling productivity improvements Abruzzese et al. [13]. Infrastructure will be required by law to be built with a plethora of embedded sensors capable of providing information on their current state in terms of structural and other functions, in order to maximize their service life against extreme environmental actions while minimizing their energy consumption and carbon footprint. The installation of IoT sensors as part of legal and structural code regulations was envisioned to obtain knowledge on the existing health of structures in order to create steps to extend their service life.

Ghosh et al. [14] investigated the potential possibilities of the IoT in the construction sector. They determined the degrees of relevance of major IoT research topics for the smart building sector. Awolusi et al. [15] investigated the use of wearable sensing devices and the IoT to improve worker safety on construction sites. They recommended using several types of sensors for physiological monitoring, ambient sensing, proximity detection, and position tracking to decrease construction dangers and improve worker safety. Jia et al. [16] and Gao et al. [17] explored the IoT technologies utilized in smart buildings and their applications. Alavi et al. [18] conducted a cutting-edge evaluation of IoT applications in the creation of smart infrastructure. Mohapatra et al. [19], using an open-source technology, investigated the construction and implementation of a low-cost IoT platform for monitoring and archiving meteorological data, namely temperature, humidity, air pressure, and dust particles in a residential area. Lee et al. [20] developed an IIoT-based health monitoring and diagnostic system to identify pump damage caused by the by-products of the atomic layer deposition process. The IoT-based health monitoring system is divided into three sections: a data gathering unit that employs six vibration sensors, a web-based monitoring unit that can monitor vibration data, and an Azure platform that looks for outliers in vibration data.

1.3 FEATURES OF DATA USED FOR HEALTH MONITORING

1.3.1 SPATIAL RESOLUTION

The spatial resolution of the data available is an important factor for analysis of any health monitoring system. It is necessary to acquire data dispersed across the structure instrumented to optimize and ensure safe operations. The DOFS used is advantageous in this regard with a high resolution. Thus, the spatial resolution of the data acquired is a unique feature necessary to evaluate the efficiency of the health monitoring of the composite structures during the in-service conditions.

1.3.2 KNOWLEDGE BASE

The data acquired should consist of data monitored from the composite system during the service environmental conditions. The knowledge of the working environments is very crucial for any health monitoring system. The features of the data to notify and optimize any operation should already have calculated failure data for a successful monitoring system.

1.3.3 Data Acquisition Rate

For any quick response action to be taken by any maintenance activity, the time stamp of the data obtained should be in real time. The data acquisition system should have acquisitions as the variations change in real time. The data monitored should be continuously updated in the server system using any protocol utilized by the health monitoring system. This ensures that the data processing algorithm should be able to handle big data growing with time.

1.4 APPLICATION OF INTERNET OF THINGS (IoT) ARCHITECTURE

Many technologies currently used in health monitoring composite structures using various techniques for data collection, processing, and manipulation are inadequate when analyzing the data metrics along with the technology drawbacks, such as imperfect data insights, low sensor consistency, and inadequacy of the monitoring system.

Instrumenting the composite structures with DOFS throughout the planned locations is an attempt to assist monitoring operations. The IoT brings together a variety of new technologies amalgamated, such as computer technology, advanced electronic architecture, and signal processing algorithms. However, due to the huge data volume, the pace of data transmission and extraction appears to be inadequate. Big data is a modern approach to solving complex problems rapidly and effectively. The technology automatically records and analyses vast amounts of data from the installed DOFS. Big data is now commonly used in a variety of industries with a major impact on the field of safety research which can effectively handle data from the composite system.

The correlation between different types of collected data cannot be effectively established, culminating in resource loss and error in accident prediction. Huge data sets can be processed effectively and quickly using cloud computing technology. This technology has the potential to provide users with useful information in a timely and secure manner. As a result, cloud computing technology can be used to process data from the composite structures instrumented with DOFS as shown in Figure 1.2.

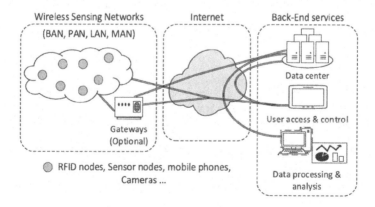

FIGURE 1.2 Architecture of an IoT system.

The connection between IoT, big data, and cloud computing technologies in health monitoring of composites structures is clarified using cloud platform and big data applications, which deliver real guarantee for the efficient implementation of IoT. To realize the development of an IoT architecture, cloud platforms and big data technology must be used. Cloud computing technology can be used to meet demand utilizing huge data sets acquired from the composite structures being monitored.

1.4.1 Technical Features of the Internet of Things (IoT) Monitoring System

By combining IoT, big data, and cloud storage technical features, a live data platform for composite structures is developed. Data collection, transmission, interpretation, and useful information production are all possible with the platform as depicted in Figure 1.3. The support layer, awareness layer, communication layer, service layer, data acquisition layer, and data storage layer make up the framework of the built platform.

The IoT uses the complex information network to collect and relay data from composite with DOFS. Based on the characteristics of data processed using cloud computing technology, an early warning system is set up and alert signals can be sent out in response to various problem scenarios. Relevant steps are introduced in a timely manner based on the platform's monitoring results to ensure the safety of monitoring system. IoT will record vast quantities of data from previous experimental operations, such as corrosive environments [21] and impact events [22]. With the cloud computing methodology, damage occurring under current conditions can be estimated based on data collected from experiments.

1.4.2 Technical Layers

1.4.2.1 Support Layer

The supporting layer mainly consists of a data system which consists of various hardware/software that facilitate the DOFS in the composite structures by providing

FIGURE 1.3 Various technical layers used in the IoT architecture.

technical assistance. The facilities would generally include a computer server system which can act as a host to enable the data server for any IoT-based health monitoring system.

1.4.2.2 Awareness Layer

The perception layer generally contains all the sensors which are employed to gather data for in real time. The parameters collected are the signal response parameters which are wavelength shift from the DOFS in the composite structures.

1.4.2.3 Communication Layer

The key to any health monitoring system-based IoT is the connectivity with the Internet or any other local network. This is the key layer which is necessary for proper functioning of the composites in the service environments. The transmission layer can consist of either wired or wireless technology connected to Internet for transmission of the data collected from the sensors to the interface in the cloud platform from the local host system. This layer should have resistance to any external perturbation as it enables the composite to be monitored for their health and optimize the maintenance activities. There are various tools in wireless technologies like Bluetooth, Wi-Fi, WAN, Zigbee or wired technologies like LAN and Ethernet used in the health monitoring system for the composite structures.

1.4.2.4 Service Layer

The historical data of composite structures, which contains all the historical knowledge based on the service environments employed, is located in the service layer. The database contains both basic and advanced information. The basic data includes information about the composite structures, its personnel and facilities, the loading conditions, geometrical structure, environmental conditions, and so on. The specialized information is inextricably linked to the safety management system. This layer is primarily responsible for receiving data sent by the communication layer. Moreover, obtained statistics is cleaned, combined, converted, and arranged at this point in order to achieve data standardization. The service layer is important for building an underground database and analyzing large amounts of data. IoT and cloud computing technology help to sustain this layer.

1.4.2.5 Data Acquisition Layer

The data acquisition layer examines simple as well as advanced data that the application layer has received and organized. Following the study, the useful data is collected and sent according to the cloud platform so that informed decisions can be taken. Mathematical analysis techniques, such as correlation analysis, should be included at this point. Correlation analysis is a mathematical technique for examining the relationship between two or more random variables. This approach can be used to investigate the factors that influence the IoT health monitoring system. In order to expand the data of the output maintenance call, research methods for different users should be carefully defined. The layer offers accurate conclusions, forecasts, and recommendations for individual users by combining the study findings with unintentional causation theories. During this stage's data processing, the service layer's internal expert knowledge base can be used.

1.4.2.6 Application Layer

The processed data is transmitted to the cloud using a LAN module, and this service is supported by 24-hour Internet connectivity. There is a summary of a full IoT digital platform. When information is accessible on the cloud server, it can be accessed quickly via the cloud host's website. The mobile app is also connected to the cloud server. As a result, the appropriate authority will take action at a local appropriate location. In order to allow for the observation of various composite positions, the information provision module will use the GIS (Maps) system to check the status of any position node in the composite being monitored in real time, and the running status of the composite structures that make up the diagnosis system in the application layer. The composite sensor data, which includes information about the surrounding environment during the service conditions, can be accessed via the data processing interface. The evaluation of the safety situation in the composite structures is evaluated. If the composite is in a high-danger scenario, concrete steps to reduce the failure rate will be recommended after the assessment. The historical data of the service layer can be accessed by the smart information catalogue in case of breakdown. The alternative backup plan management system stands at heart of this structure, and it's planned to keep the emergency rescue system running smoothly. It includes numerous emergencies, calamity, failure, and any injury catastrophe response situations. The assessment and summary module are intended to assess current issues in any disaster rescue operations.

1.5 DISCUSSION

1.5.1 Development of the Internet of Things (IoT) Architecture to Monitor the Health Under Static and Dynamic Load

According to the requirement of the end user, any required quantity of DOFS can be instrumented in the composite structures which are subjected to static-dynamic load. This aids in locating the composites' most vulnerable areas, where cracks or damages can develop in the composite structures being analyzed. Laboratory tests are carried out using an IoT-based health monitoring system. The DOFS were constituted in the sensing apparatus. The sensor response from the composite structures is used to locate any change in composite subjected to any damage. The health condition of the composite instrumented can be analyzed based on the sensor response received.

The sensor response received is sent to the host server using the Internet to the host server created. The data from the DOFS is utilized using the virtual data storage available in order to establish an IoT-based health monitoring system. The virtual data cloud server used was "ThingSpeak" which is a web service provider (REST API) that lets the user to collect and store sensor data in the cloud

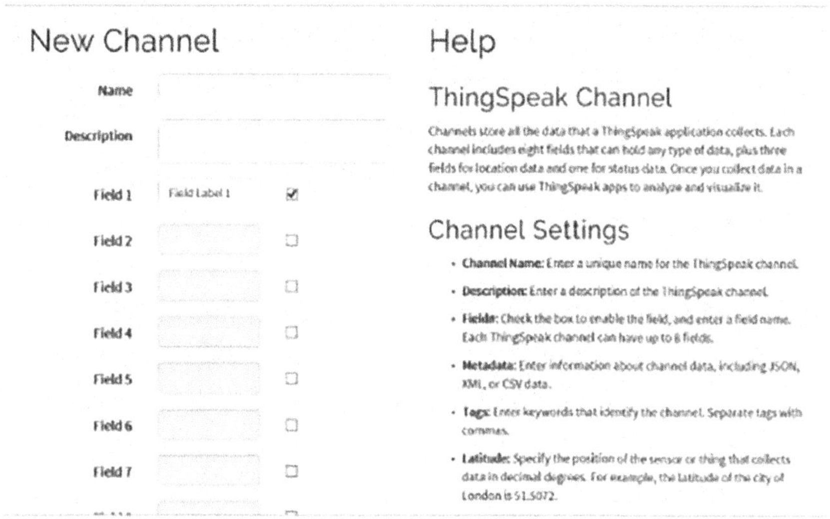

New Channel

Name	
Description	
Field 1	Field Label 1 ☑
Field 2	☐
Field 3	☐
Field 4	☐
Field 5	☐
Field 6	☐
Field 7	☐

Help

ThingSpeak Channel

Channels store all the data that a ThingSpeak application collects. Each channel includes eight fields that can hold any type of data, plus three fields for location data and one for status data. Once you collect data in a channel, you can use ThingSpeak apps to analyze and visualize it.

Channel Settings

- **Channel Name:** Enter a unique name for the ThingSpeak channel.
- **Description:** Enter a description of the ThingSpeak channel.
- **Fields:** Check the box to enable the field, and enter a field name. Each ThingSpeak channel can have up to 8 fields.
- **Metadata:** Enter information about channel data, including JSON, XML, or CSV data.
- **Tags:** Enter keywords that identify the channel. Separate tags with commas.
- **Latitude:** Specify the position of the sensor or thing that collects data in decimal degrees. For example, the Latitude of the city of London is 51.5072.

FIGURE 1.4   Collecting data using channels.

and develop IoT applications. This enables the sensor data to be accessible from anywhere in the world. The data can be secured by providing access only to the responsible authority to control the IoT-based health monitoring system to make appropriate maintenance action and schedule activities. The damage data for the delamination from the knowledge database present in the historical data can be used for analysis of the cracks or defects in the composites when subjected to static and dynamic loading scenarios. The data transmitted from sensors embedded in the composite is continuously uploaded to establish an IoT-based health monitoring system.

When the information is available at the cloud server, it can be easily accessed from the corresponding website of the cloud host. The cloud server is also linked to the mobile application. Therefore, responsible authority can check this information at anytime and anywhere as shown in Figure 1.4. The channel created assumes the data to be stored from the specific location of the DOFS in the composite structure. The sensors within the composite structure evaluate the strain data which is sent to the cloud network for analysis simultaneously.

The health of the composites at various locations when subjected to static and dynamic load can be monitored continuously using the IoT platform by using the knowledge as shown in Figure 1.5.

E-mail alerts are broadcast in response to the changes in performance of composite laminates. The email alert shown is created by running the code's threshold limits of the strain values in the range of 200–350 με identified using the knowledge base for the composite system subjected to static and dynamic load. So, if there is any strain value crossing this delamination region value, the user will receive an email letting the user know if the location needs maintenance consisting of defects/delamination as demonstrated in Figure 1.6.

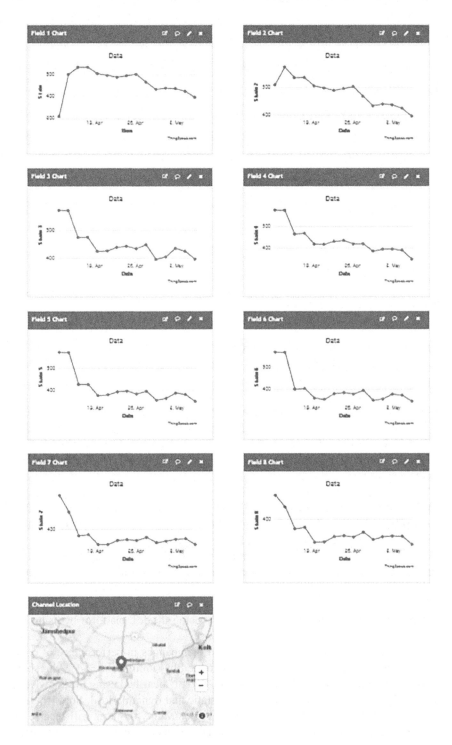

FIGURE 1.5   Monitoring static and dynamic load data using channels.

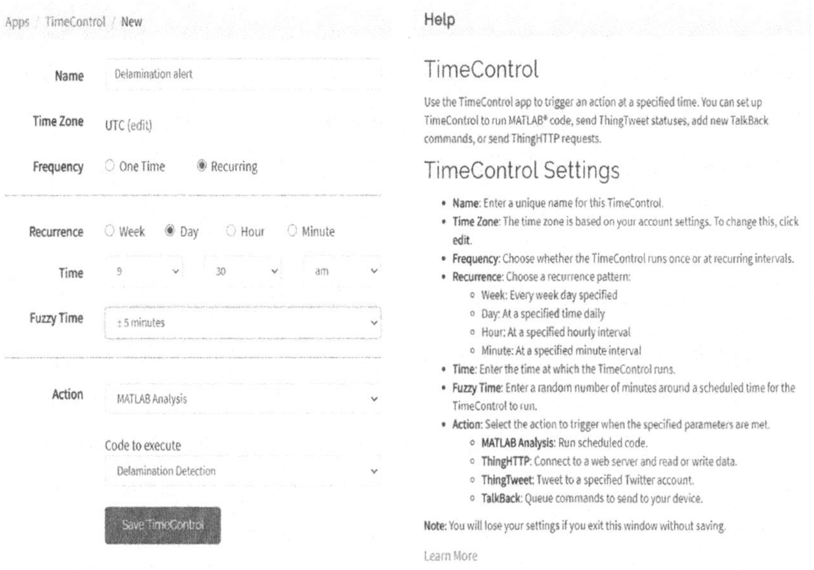

FIGURE 1.6 Sending email alert to the user instantaneously analyzing the signal response.

The email generated is shown in Figure 1.7 that allows the users to be notified about the location of the delamination in the composite system subjected to static and dynamic load.

The alerts can also be sent into various social media platforms using ThingTweet by linking the Twitter account and also WhatsApp messenger by the plug-ins available in the cloud platform by using the ThingSpeak API as shown in Figure 1.8.

Analysis and visualization tools on ThingSpeak are used for more advanced data analysis like displaying error bars, normalizing performance across analyzed from the channel data. Figure 1.9 shows the various visualization tools applied for the sensor data obtained using the DOFS. Figure 1.9 shows a (a) pie chart, (b) elucidates histogram, and (c) continuous plotting analysis, respectively.

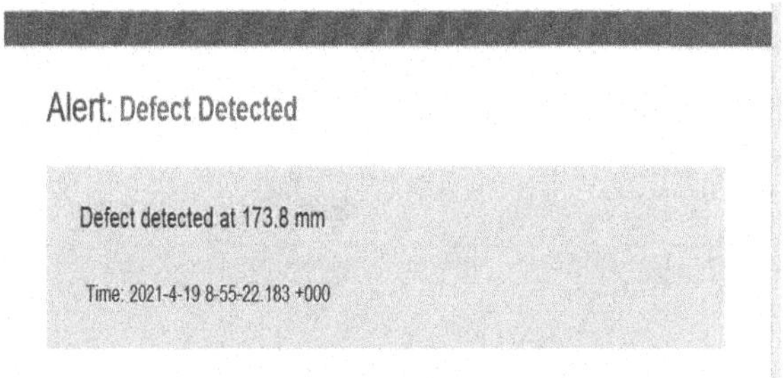

FIGURE 1.7 The email alert sent to the user to notify the defect location in the composite laminate.

FIGURE 1.8   Linking the Twitter account and tweeting the alert message for analysis.

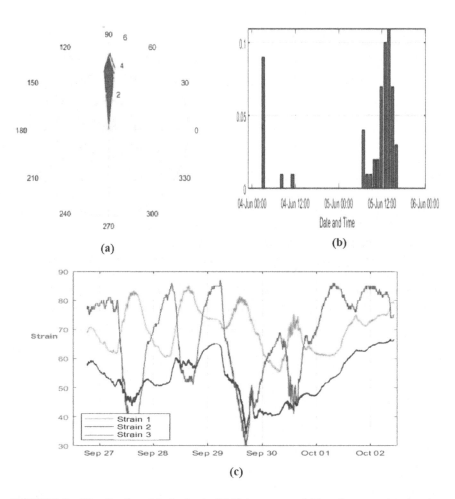

FIGURE 1.9   Visualizations (a) pie charts, (b) histograms, and (c) performance trends using the tools available to visualize in real-time analysis of static and dynamic loading in the composites.

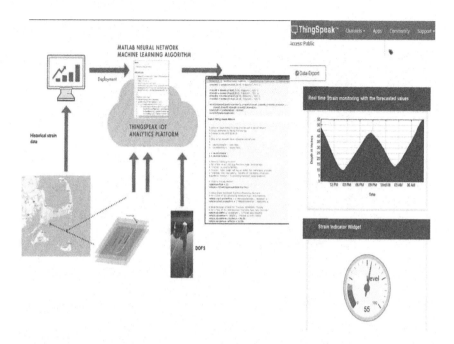

FIGURE 1.10   Machine learning algorithms for forecasting the defects using the sensor data over the locations in the composite structure.

The data acquired using the machine learning algorithms can be utilized to identify the defect location in the composite laminates which is further used to generate smart alerts. There are various machine learning algorithms that are employed to get a forecast by training the algorithm using the historical data acquired from the composite structures as shown in Figure 1.10.

The server created using the cloud platform is also linked to mobile applications to assess the health of the composite at any time and place. The mobile app is very user-friendly named "ThingView," which can be used to monitor the data from the composite structures instrumented with DOFS in real time continuously, as seen in Figure 1.11. The authority in charge can perform action based on the information for the maintenance team to act on the field. The important action plan for the future maintenance can be made using the data available.

1.5.2   DEVELOPMENT OF THE INTERNET OF THINGS (IoT) ARCHITECTURE TO MONITOR THE HEALTH UNDER IMPACT LOAD

The procedure to perform the impact monitoring operations using the cloud monitoring platform "CloudWatch" is demonstrated in Figure 1.12. The data is collected which is further analyzed using the following overview of the CloudWatch platform.

Figure 1.13 shows the overview of the dashboard used to monitor the impact data obtained using the DOFS. In impact testing, the importance of real-time monitoring

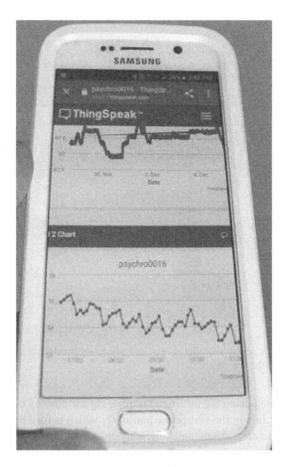

FIGURE 1.11   Mobile applications for monitoring using the sensor data in the composite structure.

along the location of the composite structure is very crucial. Thus, to evaluate the impact events, we need to add an alarm widget to the CloudWatch dashboard as seen in Figure 1.13.

The important challenge is also to make the data available be secure. Thus, the usage of the dashboard should be limited only to specific users who can handle the cloud insights. The access to perform any emergency response actions in case of

Collect   Monitor   Act   Analyze

FIGURE 1.12   Functional overview of the IoT cloud platform.

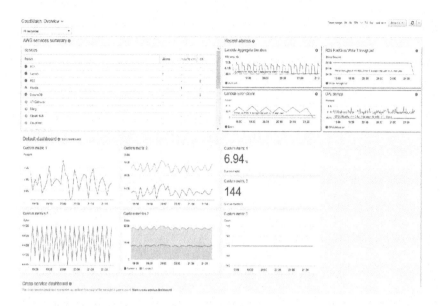

FIGURE 1.13 Default dashboard used to evaluate the sensor data.

impact events can be shared with the maintenance team in charge. CloudWatch uses the historical data which is created from the laboratory experiments from the knowledge base. The threshold limits are used to calculate a working model for predicting the values. The impact detection dashboard will utilize the pervious data to alert the user in case of low velocity impact events analyzing the data sent from the DOFS as shown in Figure 1.14.

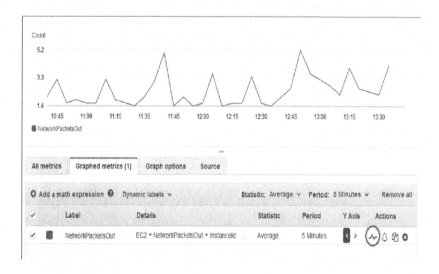

FIGURE 1.14 Historical data from DOFS study of impact event fed to set the alarm using the sensor data.

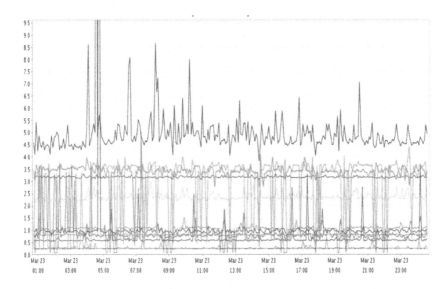

FIGURE 1.15 Detection of impact event from the DOFS data from the composites.

The data is monitored constantly can understand that the impact event can handle the data acquisition from the DOFS continuously is shown in Figure 1.15. The sudden spike marked in red is the sensor location where the impact has occurred. This can help in detecting the exact location where the impact has taken place on the composite structure.

The user using the CloudWatch can use various visualization algorithms for doing various kinds of dynamic analysis with the data being acquired. The characteristics of the sensor data can be easily mapped by using the averaging tools that aid us to comprehend the data better by choosing any of the statistics or percentiles, as shown in Figure 1.16.

FIGURE 1.16 Visualization tools for analysis of the DOFS data from the composites.

FIGURE 1.17   Averaging tools applied to evaluate the impact data from the DOFS.

After applying the averaging tool, the impact data can be easily understood as shown in Figure 1.17. The figure shows the sensor data from a single location before and after the application of the averaging tools. The difference between the two plots is clearly seen after enabling the averaging tools which filtered the noise from the raw data.

The important application of the CloudWatch is the sixth sense application used to monitor the real-time DOFS data from the composite structures in case of any impact events as shown in Figure 1.18. Thus, for each data point, depending on whether that data point value is more than the threshold limits returns the lower bound of the anomaly prediction band by setting up monitoring with the application insights.

When the user wants to perform anomaly detection using the sensor data, the CloudWatch applies machine learning algorithms to the DOFS previous data acquired from the composites to construct a model of the sensors expected values

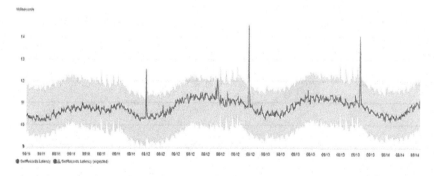

FIGURE 1.18   Anomalies predicted using the tools to evaluate the data from the DOFS.

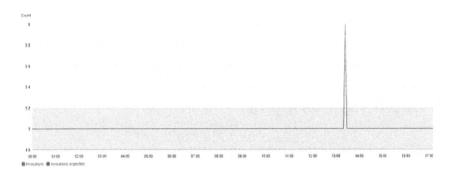

FIGURE 1.19   The model predicted result of the impact test with time.

as shown in Figure 1.19. The model considers both. The sensor trends have hourly, daily, and weekly patterns and up to two weeks of data is used to train the complete algorithm. The user can choose to exclude certain time periods from being used to train the model after which the anomaly detected using the sensor data. This way, the user can keep deployments and other unusual events out of model training, resulting in the most precise model possible.

A novel platform for IoT-based positioning of impact events of composites built on the CloudWatch, an IoT cloud platform, was presented in this section as shown in Figure 1.20. CloudWatch bridges the gap between IoT and defect location tracking by

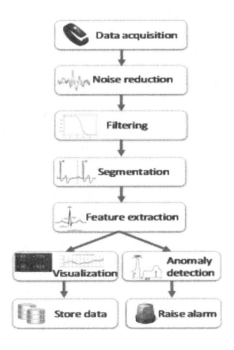

FIGURE 1.20   General overview of the IoT-based architecture for monitoring.

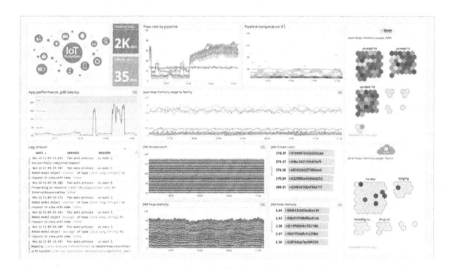

FIGURE 1.21   Data collection dashboard at various locations of the composites immersed in corrosive environments.

providing an easy-to-use and versatile tool for studying and researching composites installed with DOFS for IoT and real-time monitoring IoT applications as depicted in Figure 1.21. According to preliminary findings, monitoring of the composite structures using the CloudWatch application shows the capabilities demonstrated in a proof-of-concept implementation from the measurements of DOFS. This was used to explore the effect of device diversity and various detection algorithms on positioning accuracy in creating alarms.

1.5.3   DEVELOPMENT OF THE INTERNET OF THINGS (IoT) ARCHITECTURE TO MONITOR THE HEALTH UNDER CORROSIVE MEDIA

The entire life cycle of the IoT architecture has been shown next that depicts the methodology used to develop the IoT architecture for health monitoring in corrosive environments.

Data acquisition: For data acquisition, the DOFS were employed to continuously collect the data from the composites. The sensors were installed and the knowledge database was from the experiments performed as elucidated. The health of the composites when subjected to static and dynamic load can be assessed using the Datadog, a cloud-based IoT platform.

Advanced filtering and noise reduction: The advanced filtering option is used to evaluate the composite system monitored using DOFS using the cloud monitoring platform. The strain data at every second is being monitored using the platform where Boolean-filtered queries are run to narrow the time series data returned. Figure 1.22(a) shows the raw data and (b) is the data after the Boolean reduction algorithm was employed to filter the amount of data without losing any of the characteristic data.

FIGURE 1.22 Advanced filtering techniques to reduce the amount of (a) raw data used for (b) filtered data by reducing the noise from various sensors installed.

Classification learner: The segmentation is used to classify the sensor responses continuously based on which the detection of the important data can be monitored. In order to perform the classification analysis, the classification learner has to be input from the knowledge base with different corrosive media which is used to train the model. Without affecting system functionality, may collect granular custom metrics and events. This can be used by operators to monitor the output of connected sensors and actuators, as well as business KPIs, such as customer use. Operators can see whether their systems are fulfilling their design purpose and which sections of the stacks are causing problems in a single unified view as shown in Figure 1.23.

Feature extraction: Operators may define exactly which devices are included in a given monitor, auto-enrol new devices as they bind, and use machine learning

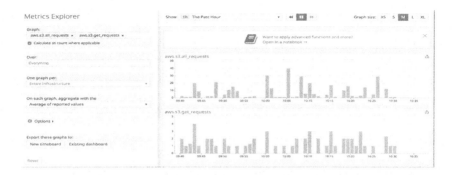

FIGURE 1.23 Classification learner to classify the data monitored from distributed optical fiber sensors (DOFS) installed.

algorithms to identify devices exhibiting abnormal behavior with versatile warnings. Individual warnings can be combined to create composite alerts that only activate if multiple conditions are met. With service-level objectives (SLO) tracking, the user can keep track of what's important to analyze. With the threshold limits known and the model trained can extract the features required as shown in Figure 1.24. Using this functionality, we can analyze the sensor data which can help us in health assessment of the composite structures.

Anomaly detection: Strain anomalies and spikes should be detected by connecting the dots between incidents and system failures in the composite structures. This is set up warnings for failed message relays and recognizes patterns in strain acquisition. The anomaly detection systems do mission-critical work, such as detecting vital failure data, sending it to the cloud, and even taking direct action. Operators can use the insights into metrics, in order to identify problems. As problems occur, operators must look at these metrics alongside hardware and firmware metrics to analyze the composites instrumented with DOFS from the corrosive conditions. The cause of the problem is dealt by analyzing the huge data acquired from the distributed sensors

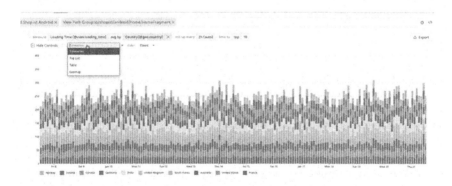

FIGURE 1.24 Feature extraction of the data acquired from the composites instrumented with the distributed optical fiber sensors (DOFS) in corrosive environment.

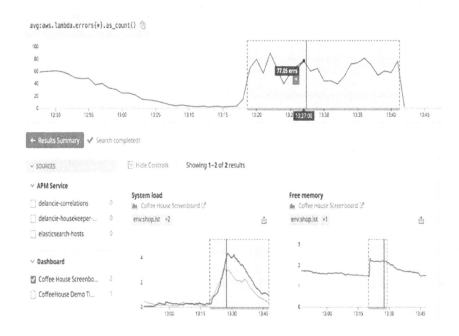

FIGURE 1.25 Fault detection from the data monitored from distributed optical fiber sensors (DOFS) installed in corrosive acids.

spread throughout the laminate. Interruptions in the flow of distributed sensor data from any one source in the corrosive environments could signify a system failure, an application-level performance issue, or a regional network issue which can be verified by maintenance team as shown in Figure 1.25.

Visualization: Time series metrics and events are produced in large quantities by DOFS sensors incorporated in composites structures. Unlike conventional sensors, these DOFS are often used in environments with intermittent connectivity and conditions that cause performance degradation. Device operators are plagued by false alarms caused by static threshold-based control, which drown out real problems. Various visualization tools are utilized to eliminate any discrepancies arising out of large data acquired as illustrated in Figure 1.26.

Store data: All major operating systems and hardware architectures are supported by Agent. The Agent was created with resource-constrained devices in mind, and it uses very few machine resources. The Agent gathers over 100 distributed sensors health indicators right out of the box using the robust APIs and client libraries for widely used languages for computers as illustrated in Figure 1.27.

Notifications: The notifications to interact with any contact or collaboration method, the data analysis can be easily done to send updates to Slack, Hangouts Chat, Microsoft Teams, and other platforms. In PagerDuty, ServiceNow, and Zendesk, the user can trigger and fix accidents. To enhance the current workflows and cause custodial actions, send alerts to web hooks. Without being fatigued, keep an eye on

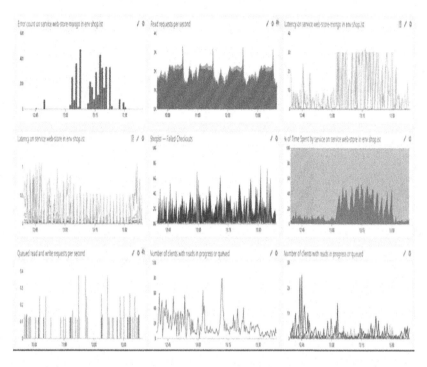

FIGURE 1.26   Visualization of senor metrics monitored from distributed optical fiber sensors (DOFS) installed.

ephemeral processes. Containers and hosts come and go. Tag-based alerts allow the user to work smarter, not harder. Tags may be used to target warnings to specific groups of hosts, containers, or other metrics. Scale up the ecosystem without missing a beat; warnings are applied to new hosts automatically. Get timely notifications about the health of the composite materials and serverless functions as illustrated in Figure 1.28.

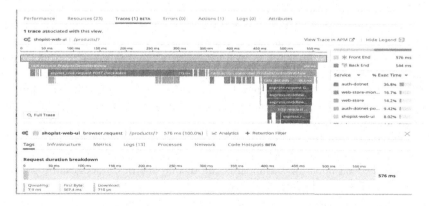

FIGURE 1.27   Storage capacity of IoT metrics using cloud platform Datadog.

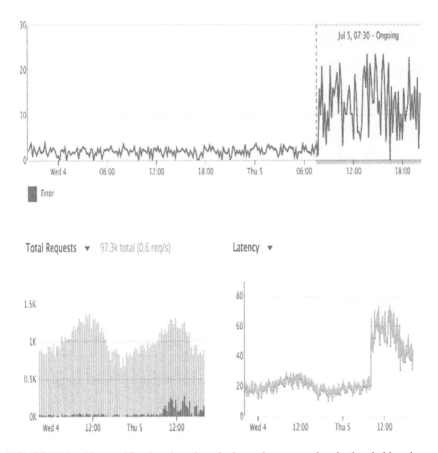

FIGURE 1.28   Alert notifications based on the incoming sensor data in threshold region.

Distribution metrics: Fleets are invariably made up of a variety of instruments with varying hardware specifications and software versions. Developers and operators can use to track all of their devices and back-end systems in a single unified network, ensuring that efficiency is optimized, problems are easily resolved, and customer satisfaction is enhanced. Companies need a detailed view of the health of their IoT fleets in aggregate, as well as the ability to drill down to troubleshoot a specific area, system model, software version, or individual device to efficiently run their IoT fleets. Monitor system fleets from top to bottom, with the ability to aggregate performance metrics by any dimension and at any degree of grunt as demonstrated in Figure 1.29.

Built-in integrations: For common IoT gateways, as well as most data processing systems, data stores, and application servers, come with pre-built integrations and out-of-the-box dashboards. Dashboards can display metrics, traces, and logs from these systems alongside data from smartphones. Operators will easily identify the source of problems, whether it's a faulty system, connection issues, or a cloud service outage as illustrated in Figure 1.30.

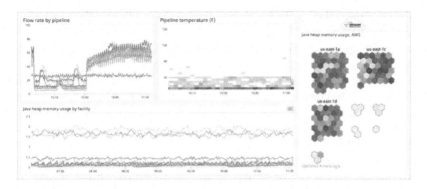

FIGURE 1.29 IoT-based sensors live at all locations of the DOFS in corrosive environment.

FIGURE 1.30 Built-in cross-platform integrations to send data and even notifications.

Notebooks: Multiple models with different processor architectures, network link types, and other hardware variations are common in IoT system fleets. Devices invariably run various versions of operating systems and program code, further complicating operations. Operators need a versatile monitoring tool that can manage this degree of heterogeneity in order to get a fleet-wide view as shown in Figure 1.31.

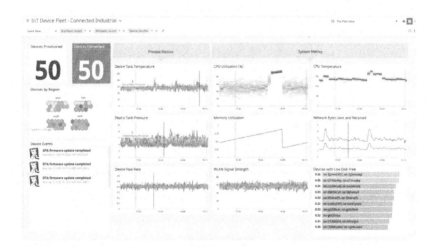

FIGURE 1.31 Segregated specific notebooks to analyze only important data.

Mobile application: With the mobile app, the user can investigate warnings in seconds. Without having to open the laptop, the user can review and evaluate vital warnings. Using just the mobile and the software on Android or iOS, the user can analyze the display warnings and anomalous data trends in seconds. Through real time, critical dashboards, the user can determine the magnitude of accidents and track the health of all the composite structures and boost monitoring by mobile application. In the coming years, billions of new IoT devices will be connected as the cost of computer hardware and sensors continues to fall and networks become faster and more widespread for IoT gives system operators an intelligent, coherent control solution so they can concentrate on developing new capabilities and pleasing consumers rather than troubleshooting device or network issues as exemplified in Figure 1.32.

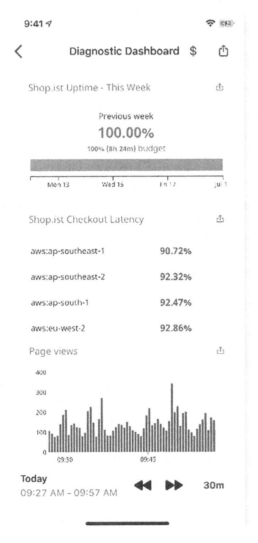

FIGURE 1.32 Mobile application to analyze innovative dashboards.

Security and compliance monitoring: IoT devices are empowering organizations across many sectors to transform their operations and deliver differentiated consumer experiences, from manufacturing to retail. Operating a large fleet of linked, geographically dispersed computers, on the other hand, presents a major challenge. Devices often operate in hostile environments and require connections to on-premise applications. IoT monitoring is fully supported by various cloud services provided by Amazon IoT, Google Cloud IoT, and Azure IoT Hub to visualize the huge data acquired from DOFS. Regardless of the source, compiles all of the operational data needed for IoT system monitoring into an easy-to-use interface for modelling, monitoring, comparison, and tracking.

1.6 CONCLUSIONS

This chapter demonstrates how an intelligent IoT structure can be used to track the health of a composite structure. Because of its potential and ability to be incorporated into any complex structure, the IoT has recently gained a lot of attention. IoT is emerging as an important technology for monitoring systems as a result of rapid growth of sensing technologies such as DOFS, as well as the integration of information technologies such as wireless communication and the Internet. The model is based on an IoT sensing system that can detect damages and sends the information to a cloud server for storage and processing. Since the data, which gives an indicator of the composite structure's health status, is accessible remotely, it can save time and money. The proposed health monitoring architecture is a real-time monitoring system that is connected to the IoT. DOFS, sensor interrogation unit, Internet network (LAN), and cloud service ThingSpeak, CloudWatch, Datadog are all part of the proposed framework. Using DOFS, the IoT monitoring system determines whether the composite structure has been damaged. The model was tested on the composite system, and it was found that the device would detect cracks/failure in the composite structure and identify possible location of problem areas. The chapter examines and presents a system for systemic health monitoring based on intelligent and accurate monitoring using IoT technologies. The technologies involved in the implementation of IoT and SHM systems, as well as data routing strategy in an IoT environment, are discussed in detail. Big data solutions are in demand as the amount of data produced by sensing devices grows more voluminous and faster than ever before. To cope with the dynamic and large amount of data obtained from DOFS embedded in composite structures, big data solutions have been implemented.

There is significant potential for IoT application in outdoor applications. The IoT revolution will expand the use of IoT in industrial composite applications, and future uses would be larger and may need incorporation into building design rules. Furthermore, advances in other study areas, such as image processing and virtual reality, will improve the IoT-based SHM of large-scale composite structures. Some questions exist about the precision of some measures and procedures for determining the structural health of composites. The early-stage maturity strategy, for example, is useful only for a few days. Furthermore, sensors may miss a variety of elements that impact the structural health of composites. To establish a match between field and laboratory data, researchers must calibrate sensors based on laboratory results and define acceptable threshold levels for monitoring parameters. Despite these limitations, IoT-based health monitoring has significant potential for usage in composite monitoring, modal-based identification techniques, concrete strength assessment, and automatic equipment creation for quality control. By permitting a quick reaction for post-disaster stabilization and repair, improvements in real-time health monitoring processes will decrease economic losses and the number of deaths caused by structural collapses. In conclusion, IoT-based health monitoring systems may improve structural safety while also meeting present demands of industrial operations.

REFERENCES

[1] Stergiou, C.; Psannis, K. E.; Kim, B.-G.; Gupta, B.: Secure integration of IoT and cloud computing. *Future Generation Computer Systems*, **78**, 964–975 (2018). https://doi.org/10.1016/j.future.2016.11.031.

[2] Carmelo Scuro, C.; Sciammarella, P. F.; Lamonaca, F.; Olivito, R. S.; Carni, D. L.: IoT for structural health monitoring. *IEEE Instrumentation & Measurement Magazine*, **21**, 4–14 (2018). https://doi.org/10.1109/MIM.2018.8573586.

[3] Du, B.; Lin, C.; Sun, L.; Zhao, Y.; Li, L.: Response prediction based on spatial-temporal deep learning model for intelligent structural health monitoring. *IEEE Internet of Things Journal*, **4662**, 1–13 (2022). https://doi.org/10.1109/JIOT.2022.3141417.

[4] Hassanalieragh, M.; Page, A.; Soyata, T.; Sharma, G.; Aktas, M.; Mateos, G.; Kantarci, B.; Andreescu, S.: Health monitoring and management using Internet-of-Things (IoT) sensing with cloud-based processing: Opportunities and challenges. Proceedings – 2015 IEEE International Conference on Services Computing, SCC 2015, 285–292 (2015). https://doi.org/10.1109/SCC.2015.47.

[5] Salem, R. M. M.; Saraya, M. S.; Ali-Eldin, A. M. T.: An industrial cloud-based IoT system for real-time monitoring and controlling of wastewater. *IEEE Access*, **10**, 6528–6540 (2022). https://doi.org/10.1109/ACCESS.2022.3141977.

[6] Belkeziz, R.; Jarir, Z.: A survey on Internet of Things coordination. Proceedings – 2016 3rd International Conference on Systems of Collaboration, SysCo 2016, **4**, 619–635 (2017). https://doi.org/10.1109/SYSCO.2016.7831328.

[7] Campelo, J. C.; Capella, J. V.; Ors, R.; Peris, M.; Bonastre, A.: IoT technologies in chemical analysis systems: Application to potassium monitoring in water. *Sensors*, **22**, 1–16 (2022). https://doi.org/10.3390/s22030842.

[8] Mishra, M.; Lourenço, P. B.; Ramana, G. V.: Structural health monitoring of civil engineering structures by using the internet of things: A review. *Journal of Building Engineering*, **48**, 103954 (2022). https://doi.org/10.1016/j.jobe.2021.103954.

[9] Rewatkar, P.; Nath, D.; Kumar, P. S.; Suss, M. E.; Goel, S.: Internet of Things enabled environmental condition monitoring driven by laser ablated reduced graphene oxide based Al-air fuel cell. *Journal of Power Sources*, **521** (2022).

[10] Haghnegahdar, L.; Joshi, S. S.; Dahotre, N. B.: From IoT-based cloud manufacturing approach to intelligent additive manufacturing: Industrial Internet of Things – An overview. *International Journal of Advanced Manufacturing Technology*, **119**, 1461–1478 (2022). https://doi.org/10.1007/s00170-021-08436-x.

[11] Mohapatra, A. G.; Talukdar, J.; Mishra, T. C.; Anand, S.; Jaiswal, A.; Khanna, A.; Gupta, D.: Fiber Bragg grating sensors driven structural health monitoring by using multimedia-enabled IOT and big data technology. *Multimedia Tools and Applications* (2022). https://doi.org/10.1007/s11042-021-11565-w.

[12] Hernández-Morales, C. A.; Luna-Rivera, J. M.; Perez-Jimenez, R.: Design and deployment of a practical IoT-based monitoring system for protected cultivations. *Computer Communications*, **186**, 51–64 (2022). https://doi.org/10.1016/j.comcom.2022.01.009.

[13] Abruzzese, D.; Micheletti, A.; Tiero, A.; Cosentino, M.; Forconi, D.; Grizzi, G.; Scarano, G.; Vuth, S.; Abiuso, P.: IoT sensors for modern structural health monitoring. A new frontier. *Procedia Structural Integrity*, **25**, 378–385 (2020). https://doi.org/10.1016/j.prostr.2020.04.043.

[14] Ghosh, A.; Edwards, D. J.; Hosseini, M. R.: Patterns and trends in Internet of Things (IoT) research: Future applications in the construction industry. *Engineering, Construction and Architectural Management*, **28**, 457–481 (2021). https://doi.org/10.1108/ECAM-04-2020-0271.

[15] Awolusi, I.; Nnaji, C.; Marks, E.; Hallowell, M.: Enhancing construction safety monitoring through the application of Internet of Things and wearable sensing devices: A review. Computing in Civil Engineering 2019: Data, Sensing, and Analytics – Selected Papers from the ASCE International Conference on Computing in Civil Engineering 2019, 530–538 (2019). https://doi.org/10.1061/9780784482438.067.

[16] Jia, M.; Komeily, A.; Wang, Y.; Srinivasan, R. S.: Adopting Internet of Things for the development of smart buildings: A review of enabling technologies and applications. *Automation in Construction*, **101**, 111–126 (2019). https://doi.org/10.1016/j.autcon.2019.01.023.

[17] Gao, X.; Pishdad-Bozorgi, P.; Shelden, D. R.; Tang, S.: Internet of Things enabled data acquisition framework for smart building applications. *Journal of Construction Engineering and Management*, **147**, 04020169 (2021). https://doi.org/10.1061/(asce)co.1943-7862.0001983.

[18] Alavi, A. H.; Jiao, P.; Buttlar, W. G.; Lajnef, N.: Internet of Things-enabled smart cities: State-of-the-art and future trends. *Measurement: Journal of the International Measurement Confederation*, **129**, 589–606 (2018). https://doi.org/10.1016/j.measurement.2018.07.067.

[19] Mohapatra, D.; Subudhi, B.: Development of a cost effective IoT-based weather monitoring system. *IEEE Consumer Electronics Magazine*, **2248** (2022). https://doi.org/10.1109/MCE.2021.3136833.

[20] Lee, Y.; Kim, C.; Hong, S. J.: Industrial Internet of Things for condition monitoring and diagnosis of dry vacuum pumps in atomic layer deposition equipment. *Electronics (Switzerland)*, **11** (2022). https://doi.org/10.3390/electronics11030375.

[21] Padarthi, Y.; Mohanta, S.; Gupta, J.; Neogi, S.: Assessment of transport kinetics and chemo-mechanical properties of gf/epoxy composite under long term exposure to sulphuric acid. *Polymer Degradation and Stability*, **183** (2021). https://doi.org/10.1016/j.polymdegradstab.2020.109436.

[22] Padarthi, Y., Mohanta, S., Gupta, J.; Neogi, S.: Quantification of swelling stress induced mechanical property reduction of glass fiber/epoxy composites immersed in aqueous 10% sulphuric acid by instrumenting with Distributed Optical Fiber Sensors. *Fibers and Polymers*, **23**, 212–221 (2022). https://doi.org/10.1007/s12221-021-0317-2.

2 Computer-Aided Ergonomic Design and Assembly of a Domestic String Hopper Machine

C. Vigneswaran, G. Madhan Mohan,
M. Vignesh Raja, E. Selva Vignesh
PSG College of Technology, Coimbatore, Tamil Nadu,
India

CONTENTS

2.1 INTRODUCTION

Indian culture has a wide variety of traditional foods. In India, houses still depend on traditional conventional machine tools for making certain regional dishes. Among them, 'sev,' a traditional South Indian food cooked with flour strings, is one of the widely consumed dishes. The types of equipment available to prepare these foods for domestic purposes were human-powered and are subjecting the user to musculoskeletal disorders.

The few available pieces of equipment are based on mechanisms, such as screw rod and lever type, and are suitable for batch production in small-scale industries [1, 2]. The need for electric-powered string hopper machines for domestic purposes was unaddressed to date and the focus of this chapter is to address the issue on how to reduce user fatigue. The new approach was focused on the design and development of an electric string hopper machine that is suitable for domestic purposes.

The proposed method can be used to make flour strings continuously without interruption. The new design consists of a feeding hopper, screw compression mechanism, die for extrusion, electric motor, and a base.

The embodiment design was carried out followed by a detailed design with CREO 5.0. Rapid Upper Limb Assessment (RULA) analysis was carried out using CREO 5.0 with 50th percentile digital human model. The scores were found to be 2 for both arms and are acceptable. The prototype was tested in a real-time environment for the extrusion of steam-cooked rice flour and the flour strings were produced [3]. The extrusion rate was found to be 2 kg/hour with a die hole size of 5 mm diameter. The product developed was proven to be ergonomically safe and are suitable for domestic use.

The main objective of the work is to redesign a power assisted domestic string hopper-making machine to minimize the human effort. The existing machine is manual assisted and that is outdated which requires more tedious work effort of the human while preparing the dish. Hence, an automatic string hopper machine for a domestic purposes has to be developed.

Machines available in the market are suitable only for small- and medium-scale industries. A few available equipment are based on mechanisms, such as screw rod and lever type, and are suitable for batch production only in small-scale industries [1]. The problem with these mechanisms is that the extrusion cylinders need to be refilled after each batch, and also, they are human-powered. Another study focused on the development of a string hopper machine with triple cylinders to reduce the refilling time [2], but again the capacity is suitable only for industrial applications. The domestic application-based string hopper machines are not available in the market. The various existing mechanical models available in the market have not yet solved this issue. From the literature survey, the authors have framed the problem statement and further found a computer-aided solution by developing a new product.

2.2 METHODOLOGY

The methodology involves problem definition, in which the problem was identified and defined. This chapter is focused on addressing the issue of the need for an electrified model of domestic string hopper machine, which reduces the ergonomic risks of the users. The next step was the literature survey. In this step, it was identified that there already exist electric string hopper machines for industrial purposes, but the capacities and the cost are too high to accommodate them for domestic use. Hence, it is clear that there is a need for the development of an electric-powered domestic string hopper machine to address the problem defined. To begin with the development of a solution, few concepts were generated. The relative advantages and disadvantages were compared, and one was selected based on the pug matrix. Then for the selected concept, the embodiment design was carried out. Using CREO 5.0, a detailed 3D design model was generated and the cost estimation for the prototype was done. The cost estimation details are discussed thereafter. Then the RULA analysis was done using CREO 5.0, and the procedure and result of the analysis are discussed in detail. A prototype was made as per the design and finally tested under a real-time environment, and the results are discussed later in this chapter.

2.3 PRODUCT DESIGN FLOW

The product consists of 12 parts. The parts include base, cast iron body, compression screw, extrusion die, end cover, helical gear, pinion, a 12V DC motor, M10 bolts, M10 nuts, M20 bolts, and M8 bolts as tabulated. The base is made up of mild steel plates of 5 mm thickness. The function of the base is to support all other parts of the machine, and it rests on top of a table. The mild steel has good weldability; therefore, it best suits the purpose. The cast iron body, compression screw, and end cover are made up of cast iron. The body parts and end cover have a complex shape; hence, the selection of casting method is the most economical one.

Further, to make it suitable for food processing, an enamel coating was provided over the surface area of the parts. The extrusion die was made up of stainless steel which is a food-grade material. The extrusion die is replaceable, and it can be made as per the required shape as intended by the user. The helical gear and pinion are made up of mild steel. The purpose of the gear is to increase the torque and reduce the speed. Helical gears have a module of 1.3, and the number of teeth is 86 and 26, correspondingly. The helix angle and the pressure angle of the gears are 20 degrees. The gear tooth is case-hardened to prevent wearing of the gear tooth. The internal bores of the gears are to be machined as per the geometry of their corresponding mating shafts. The bolts and nuts are made up of low-carbon steel. The DC motor with gearbox assembly was procured for assembly. The DC motor has a voltage rating of 12 volts and the current rating is 2.5 ampere. The rated output torque of the motor is 18 Nm at 13.5 V. The operating temperature range for the motor is 20–90 degree Celsius. Appropriate quantities are tabulated corresponding to the parts. For a detailed requirement of bill of materials, refer to Table 2.1. For the ease of product visualization, the exploded view of computer-aided model is shown in Figure 2.1.

TABLE 2.1
Bill of Materials for the Newly Developed Model

S. No.	Description	Material	QTY
1	Base	Mild steel	1
2	CI body	CI	1
3	Compression screw	CI	1
4	Extrusion die	Stainless steel	1
5	End cover	CI	1
6	Helical gear	Mild steel	1
7	Pinion	Mild steel	1
8	12V DC motor with gearbox	–	1
9	M10 bolt	Low carbon steel	4
10	M10 nut	Low carbon steel	5
11	M20 nut	Low carbon steel	1
12	M8 bolt	Low carbon steel	3

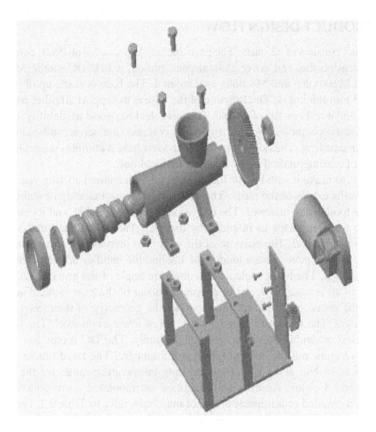

FIGURE 2.1 Exploded view of computer-aided machine assembly.

2.3.1 Cost Estimations

The concept of frugal innovation has been considered while selecting the subassembly. The cast iron body, compression screw, extrusion die, and end cover consist
of the subassembly. The aforementioned subassembly is a meat grinder. The actual
assembly consists of a grinding blade between the extrusion die and the compression
screw at the extrusion end. This grinding blade is removed and the rest of the assembly is used as it is. The cast iron parts are already coated and are of food-processing
grade.

Parts like M10 bolts, M10 nuts, M20 nuts, and M8 bolts are procured. The helical
gear, pinion, and 12V DC motor are also procured. The base is fabricated with mild
steel plates with shielded metal arc welding. The helical gear and pinion boreholes
are machined to accommodate their corresponding shaft geometry [4]. The machining cost is applicable as per the local standard. The parts and subassemblies are then
assembled. Hence, fabrication, machining, and assembly charges are applicable as
per the local standard. For a detailed view, refer to Table 2.2.

TABLE 2.2

Cost Estimation for a Prototype Made

S. No.	Part/subassembly	Cost in Rupees
1	Base	150
2	CI body	1300
3	Compression screw	
4	Extrusion die	
5	End cover	
6	Helical gear	150
7	Pinion	100
8	12V DC motor with gearbox and 230V AC to 12V DC adaptor	700
9	M10 bolt	50
10	M10 nut	50
11	M20 nut	15
12	M8 bolt	15
	Machining and fabrication charges	300
	Total cost in rupees	2830

2.4 WORKING MECHANISM AND ANALYSIS

The product would be placed at a height of 80 centimeters from the floor level. The person would stand in front of the machine to operate it. A 12V DC power is connected to the motor through a switch. The DC motor has a voltage rating of 12 volts and the current rating is 2.5 ampere. The rated output torque of the motor is 18 Nm at 13.5V. The operating temperature range for the motor is 20–90 degree Celsius. Once the machine is turned on, the pinion gear receives power from the motor through the gearbox. The pinion in turn rotates the helical gear. As the gear is coupled with a compression screw, the compression screw rotates. The purpose of the gear is to increase the torque and reduce the speed. When the user feeds the food material to be extruded into the feeding hopper, the compression screw starts compressing the food material; food must be continuously fed into the hopper [5]. Once the compression volumetric space of the machine is loaded, the extrusion begins. The extruded string can be collected in a vessel. The extrusion happens until the user stops feeding the hopper.

2.4.1 ERGONOMIC ANALYSIS

The computer-aided 3D modeling was performed using CREO 5.0. The parts were designed using the top-down approach. In the CREO assembly, the complete assembly of the machine is imported. A 50th percentile manikin is imported into the CREO assembly. The product is assumed to be placed at a height of 80 centimeters from the floor level. The manikin is supposed to stand in front of the machine and operate it. In this regard, a hypothetical plan was constructed wherein the imported manikin was made to stand at a distance of 80 centimeters to act as a datum. The manikin was controlled by making it stand up on the floor plan and facing orientation with other available plans. Hands and head positions are adjusted to represent the actual working posture of the user. The user's arm would neither be supported nor would the user lean. The user's arm was assumed to be in a standing position and would not be working across the midline or outside the body. The user's trunk as well as his/her legs and feet would be well supported and balanced. The posture would not be static. The user's load should not exceed more than 2 kg.

The inputs for computation of the RULA score for the right hand were given as discussed previously. In this condition, the RULA score for the right hand is evaluated. The RULA score was found to be 2, which is acceptable. For the left hand, the RULA score was evaluated with the following inputs: the user's arm would neither be supported nor would the user lean. The user's arm would not be working across the midline or outside the body. The user's trunk would be well supported.

The user, if needed, can have a seat of suitable height, but for this analysis, the user is assumed to be in a standing position. The user's legs and feet would be well supported and balanced. The posture would not be static. The user's load should not exceed more than 2 kg. Figure 2.2 illustrates the RULA score for the left arm that is also found to be 2, which is acceptable. It is proven that the machine is ergonomically safe [6].

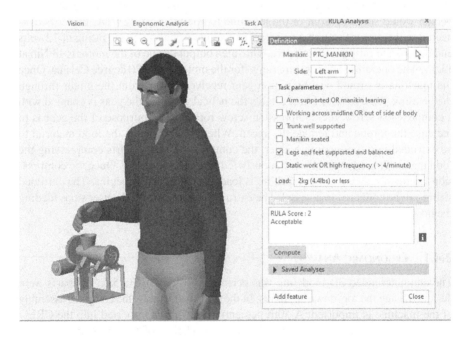

FIGURE 2.2 RULA analysis for left arm.

2.5 PROTOTYPING AND TESTING

The part drawings are generated before prototyping. A 5 mm thick plate is cut into 200 mm × 200 mm for fabrication of the machine base. The markings were made on this plate to indicate the region where the other pieces need to be welded. The four pieces 150 mm × 15 mm were cut and welded to the base, in such a manner that the four pieces stand perpendicular to the base plate, which acts as four columns. Another piece of 180 mm × 60 mm was cut and welded to the base at the appropriate location, which is to hold the motor subassembly. Further, another two pieces of size 160 mm × 15 mm were cut and a 10 mm diameter slot of width 15 mm were drilled at four points is then welded to the previously welded columns. The inner bore of the mild steel pieces was machined to concern M20 thread. An M20 nut was tightened against the gear to constrain them axially to M8 bolts. In the pinion bore, internal straight knurling was done. The output shaft of the motor gearbox was threaded at the external end and straight knurled at the internal end. The pinion bore has been employed properly and the bolts are tightened finally. The functional prototype is shown in Figure 2.3. A 12V DC power was generated from a 230V AC power supply. The adaptor provides a power supply of 12 volt and 2.5 ampere.

The prototype was tested using the semi-solid rice flour, commonly used in households. The prototype was tested in a real-time environment for the extrusion of steam-cooked rice flour. The steam-cooked flour was made into small momo-shaped pieces and it was fed continuously into the hopper. For the first few seconds, the compression volume inside the body gets loaded. After a few seconds, once the machine gets loaded, then the flour strings start getting extruded. A vessel has been kept in front of the extrusion end of the machine to collect the extruded strings. The extrusion rate was found to be 2 kg/hour with a die hole size of 5 mm in diameter. Thus, the product developed shown in Figure 2.4. was proven to be ergonomically safe and are suitable for domestic use [7].

FIGURE 2.3 Functional prototype of the machine.

FIGURE 2.4   Real-time prototype testing.

2.6   RESULT AND DISCUSSION

Ergonomic assessment (RULA analysis) was performed using CREO 5.0 software. The digital manikin had been developed and imported into the software. The 50th percentile manikin has been chosen for the assessment. The foot of the manikin has been constrained with the plane representing the floor. The hand of the manikin has been positioned as per the working posture. The evaluated RULA score for the right arm has been found to be 2. As per the RULA action-level chart, the RULA score of 2 is an acceptable posture. The RULA analysis has been repeated for the left arm. The RULA score for both arms was evaluated, and the scores were acceptable. The prototype was tested in a real-time environment for the extrusion of steam-cooked rice flour and the flour strings were produced [8]. The real posture was evaluated for RULA score. Also, the RULA score for both the right and left hands of the operator has been found to be 2. The digital mockup and the experimental values are found to be the same and, hence it has been verified both digitally and experimentally that the machine was ergonomically pleasing. The capacity of the machine was tested further; the flour has been fed into the hopper and the rate of extrusion was calculated. The extrusion rate was found to be 2 kg/hour with a die hole size of 5 mm diameter. The extrusion was repeated for few more times and it was found to be consistent.

2.7   CONCLUSION

Thus, the design and development of the domestic string hopper machine were successfully made and the design was validated for ergonomic risks by RULA analysis [9]. The RULA score for both the right and left arms was found to be 2. The RULA analysis was repeated for both the arms and found acceptable. Therefore, it has been proved that the machine is ergonomically safe. The prototype has been tested in a real-time environment for the extrusion of steam-cooked rice flour and the flour strings were produced. The extrusion rate was found to be 2 kg/hour with a die holes size of 5 mm diameter. The product thus developed for domestic purpose was ergonomically pleasing and would prevent musculoskeletal disorders [10].

REFERENCES

1. Subramani, Deepak, Manonmani Kumaraguruparaswami, Harshni Muthusamy, Sangeetha Arunachalam, and Gowthami Shanmugam. (2022). Formulation and quality evaluation of quinoa enriched ready to cook string hoppers (Indian traditional noodles). *Journal of Culinary Science & Technology*: 1–20.
2. Bhandari, V. B. (2017). *Design of Machine Elements*. India: McGraw-Hill Education.
3. Rabbani, A., and S. Ahmed (2020). Ergonomic analysis of material handling for a residential building at Rourkela. *Journal of the Institution of Engineers (India): Series A*, 101(4): 689–699.
4. Oakman, Jodi, Victoria Weale, Natasha Kinsman, Ha Nguyen, and Rwth Stuckey. (2022). Workplace physical and psychosocial hazards: A systematic review of evidence informed hazard identification tools. *Applied Ergonomics*, 100: 103614.
5. Mital, Anil, Andrew S. Nicholson, and Moh M. Ayoub. (2017). *A Guide to Manual Materials Handling*. London: CRC Press.
6. Isa, Munawwarah Solihah Muhammad, Nurhidayah Omar, Ahmad Faizal Salleh, and Mohammad Shahril Salim. (2021). A literature review on occupational musculoskeletal disorder (MSD) among industrial workers in Malaysia. *Intelligent Manufacturing and Mechatronics*, 1069–1079. doi: 10.1007/978-981-16-0866-7_95.
7. Perera, Philip (2019). Modified version of Semi-Automatic String Hopper machine. Symposium NERD Center.
8. Madhwani, Kishore P., and P. K. Nag. (2017). Effective office ergonomics awareness: Experiences from global corporates. *Indian Journal of Occupational and Environmental Medicine*, 21(2): 77.
9. Jain, Rahul, Makkhan Lal Meena, Govind Sharan Dangayach, and Awadhesh Kumar Bhardwaj. (2018). Association of risk factors with musculoskeletal disorders in manual-working farmers. *Archives of Environmental & Occupational Health*, 73(1): 19–28.
10. Sharma, Neelesh K., Mayank Tiwari, Atul Thakur, and Anindya K. Ganguli. (2022). An ergonomic study of the cleaning workers to identify the prevalence of musculoskeletal disorders in India. In *Technology Enabled Ergonomic Design*, pp. 33–46. Singapore: Springer.

Section II

Automation and Optimization

3 Application of Grey Wolf Optimizer Algorithm in Electrochemical Machining Process Optimization

S. Ghosh
Jadavpur University, Kolkata, West Bengal, India

B. Acherjee
Birla Institute of Technology: Mesra, Ranchi, Jharkhand, India

A.S. Kuar
Jadavpur University, Kolkata, West Bengal, India

CONTENTS

3.1 INTRODUCTION

Electrochemical machining (ECM) is a high-precision machining technique. This machining technique is based on the basic electrolysis concept, which includes passing a direct electric current through an electrolyte, resulting in chemical reactions at electrodes and substrate decomposition. In electrolysis, material is liberated from the anode and deposited in the cathode; in ECM, the workpiece serves as the anode

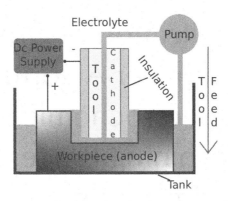

FIGURE 3.1   Scheme of an ECM system.

and as current flows, materials are removed from it for machining. The material is prevented from settling on the cathode (the tool) by a continuous electrolyte flow across the tool–workpiece gap, and a constant gap is maintained by giving a regulated down feed to the tool (Figure 3.1). ECM can be used to manufacture any form on the workpiece since it generates a direct mirror copy of the tool on it [1, 2]. ECM is used for cutting, drilling, trepanning, deburring, micromachining, profiling, and contouring in a variety of industrial applications, including drilling jet engine turbine blades and creating complicated concave curvature of steam turbine blades within tight tolerances. Because ECM does not produce stresses on the workpiece surface, it is frequently used to machine the most delicate and toughest materials in thin-walled geometries. ECM has become a viable solution for machining difficult-to-machine materials and creating complex shapes due to its numerous benefits, which include negligible tool wear, exceptional machining performance, and non-dependency on mechanical properties of the material, which is not possible in the conventional method [2, 3]. Though ECM has several advantages, frequent issues with ECM include effective tool-design functioning, electrolyte management, and metal hydroxide sludge disposal.

A number of experimental and numerical research works on ECM process mechanics, process simulation, and parametric studies are conducted by several researchers to better understand the process for yielding better performance and throughput [2–12]. These research works reveal that electrochemical phenomena and material removal rate may be regulated by ECM process variables including current, voltage, type and concentration of electrolyte, electrolyte flow rate, tool feed rate, tool material, and inter-electrode gap. These parameters also determine the machining quality, which is characterized by machining quality features, such as surface finish, circularity, overcut, tapering of the wall, and kerf width. ECM process parameters must be carefully controlled to achieve the necessary machining performance and quality. However, this is challenging since some of the process responses are contradictory, necessitating a precise trade-off between them in order to achieve multiple goals at the same time. Engineers and decision-makers are tackling such challenges using contemporary optimization techniques in order to get better outcomes while working with available resources. Researchers are constantly working to develop, analyze, and validate optimization strategies and apply them to many fields of engineering, including manufacturing. Several efforts are performed in this line to optimize the ECM parameters to obtain the required machining qualities. Rao et al. [13] optimized the ECM parameters to enhance dimensional accuracies, tool life, material removal rate (MRR), and ECM process operating cost. Asokan et al. [14] determined optimal ECM parameters including voltage, current, inter-electrode distance, and electrolyte flow rate by employing regression model and artificial neural network (ANN) to improve MRR and surface finish in ECM of hardened steels. Samanta et al. [15] employed an artificial bee colony (ABC) algorithm in search of optimal process conditions for a number of nonconventional machining techniques, which include ECM. Rao and Kalyankar [16] employed a teaching learning-based optimization (TLBO) algorithm for identifying optimum ECM parameters. Das et al. [17] examined the ECM of EN 31 steel and optimized process parameters to enhance MRR and surface finish utilizing ABC algorithm. Acherjee et al. [18] investigated the ECM process for optimizing ECM parameters to enhance the MRR and surface finish using flower pollination algorithm (FPA). Sekar et al. [19] investigated the nanoparticle-suspended electrolyte-enabled ECM process and optimized the process parameters by employing the artificial fish swarm algorithm (AFSA).

The grey wolf optimizer (GWO) is a unique swarm intelligence-based metaheuristic algorithm. It is frequently adopted for a broad range of optimization issues because of its benefits compared to alternative swarm-based algorithms: it has very few variables and does not need derivation knowledge in the initial search. It is also faster, simple, dynamic, and configurable, with the unique ability to strike the proper balance between innovation and utilization throughout the process, resulting in favorable convergence [20]. This approach is driven by the societal hierarchies and hunting skills of grey wolves. GWO has recently been widely used in a variety of disciplines ranging from artificial intelligence to engineering problems, biomedical sectors, and communication engineering. GWO is used by Li and Wang [21] to optimize the settings of PI controller in a closed-loop condenser pressure control system. The trials revealed that GWO outperformed other optimization algorithms like the PSO and genetic algorithm (GA). Yada et al. [22] sought to tweak the parameters of a standard PID controller to raise a magnetic levitation device's metal ball. GWO engineering tuning system surpasses the classic Ziegler–Nichols (ZN) tuning system. GWO was utilized by Das et al. [23] for optimization of the PID controller settings for speed modulation in a DC motor, outperforming PSO, ABC, and ZN. Kumar and Pant [24] used the GWO to solve system reliability optimization issues involving an intricate bridge and a space capsule life-support component. When the GWO algorithm results are compared to those of the simulated annealing (SA), ACO, PSO, and cuckoo search (CS) algorithms, it is found that GWO produced excellent results. GWO was utilized by Khalilpourazari et al. [25] for optimizing the variables of a nonlinear multi-constrained mathematical model for multi-pass milling process. In terms of generating better and viable solutions, the proposed algorithm outperforms existing optimization approaches in the literature along with metaheuristic algorithms like multi-verse optimizer (MVO) and dragonfly algorithm (DA). Komakia and Kayvanfar [26] used GWO in an assembly flow-shop scheduling problem for determining the ideal work sequencing to process the project in the shortest feasible time frame. Rameshkumar et al. [27] present a GWO-based optimization method for improving a thermal power plant's power unit commitment schedule. A comparison study is carried out, and it is found that the proposed GWO-based strategy outperforms a number of existing techniques.

In this work, the GWO algorithm is used for optimization of the ECM parameters to improve MRR and machining quality as measured by surface roughness (Ra) and radial overcut (ROC). The GWO algorithm is employed for identifying optimal process parameter settings for single-objective and multi-objective optimization of the ECM process. Two separate cases of ECM are considered to apply and evaluate the GWO algorithm for its efficiency with regards to optimal results, convergence, and accuracy. Finally, GWO algorithm's performance is contrasted with Gauss Jordan, ABC, FPA, and TLBO algorithms. The results obtained by employing the GWO algorithm demonstrate its applicability and effectiveness in improving ECM operational efficiency.

3.2 GREY WOLF OPTIMIZER ALGORITHM

Gray wolf family groupings are hierarchical, and social coordination among them indicates higher level of intelligence, which assists in strategic hunting planning. The GWO algorithm is based on tracking movements of wolves during the hunting phase. The following are the major phases in the grey wolf hunting method [20, 21]:

1. searching for prey
2. following the prey, chasing it down, and approaching it
3. pursuing, chasing, and pestering the prey until it comes to a halt
4. encircling and attacking the prey

In a grey wolf pack, four sorts of wolves are reported based on their hierarchical rank: alpha (α) wolf > beta (β) wolf > delta (δ) wolf > omega (ω) wolf. The α-wolf is in charge of making choices regarding hunting, sleeping, where to sleep, when to wake up, and other activities. The second rung of the grey wolf hierarchy is β-wolf, which assists the α-wolf in decision-making and other tasks, as well as providing feedback to α-wolf. The δ-wolf comes after the β-wolf and the ω-wolf comes at the bottom of the wolf's hierarchical order. The GWO algorithm uses α-, β-, and δ-wolves to lead the searching and these three wolves are pursued by the ω-wolves. The hunting behavior of grey wolves may be quantitatively explained using succeeding equations [28–30]:

Let us split the complete surroundings in which the wolf will pursue the prey into two- or three-dimensions vector space. Each and every conceivable location on the matrix can be a wolf position, and, therefore, a solution. Let the prey to stay at a location Xp while any wolf at any time instant t is at the position $X(t)$. As a result, their distance (D) is

$$D = \left| CX_p - X(t) \right| \tag{3.1}$$

$$X(t+1) = \left| X_p(t) - A.D \right| \tag{3.2}$$

where, X represents the grey wolf's location, Xp represents the prey's location, t is the current iteration, A and C are the coefficient vectors given as,

$$A = 2ar_1 - a$$

$$B = 2r_2$$

where, r_1, r_2 [0,1] are the random vectors; the components of a are progressively lowered from 2 to 0 with iterations.

To mathematically replicate grey wolf hunting behavior, it is believed that α (best), β, and δ have greater awareness of probable prey locations. The following equations are presented to represent the grey wolf hunting strategy.

$$D_\alpha = |C_1 X_\alpha - X(t)|, \quad D_\beta = |C_2 X_\beta - X(t)|, \quad D_\delta = |C_3 X_\delta - X(t)| \tag{3.3}$$

$$X_1 = |X_\alpha - A_1 D_\alpha|, \quad X_2 = |X_\beta - A_2 D_\beta|, \quad X_3 = |X_\delta - A_3 D_\delta| \tag{3.4}$$

$$X(t+1) = |X_1 + X_2 + X_3|/3 \tag{3.5}$$

A search agent in an n-dimensional search space can use these equations to adjust its location based on α, β, and δ.

Grey wolves, as previously indicated, end the hunt by attacking the prey once it has stopped running. To numerically describe reaching the target, one decreases the value of a. It is also worth noting that the fluctuation range of A is likewise deceased by a. The condition $|A| < 1$ causes the wolves to attack the prey. The values of $|A|$ and $|C|$ are also important in selecting and emphasizing the prey. $|A| > 1$ leads the grey wolves to depart from their prey in quest of a more fit prey. C is another GWO factor that assigns randomized weights to prey in emphasizing ($C > 1$) or de-emphasizing ($C < 1$) their value in influencing the distance in Equation (3.1). The flowchart in Figure 3.2 depicts the processes of a GWO algorithm.

3.3   OPTIMIZATION OF ECM PROCESS

The GWO algorithm is employed to optimize the ECM process on a single- and multi-objective basis. The GWO algorithm is implemented in two ECM case studies for finding the optimal process conditions to obtain the intended process performance and quality. The objective functions used to solve using the GWO algorithm are the regression models developed to correlate the ECM parameters with the anticipated response parameters based on experimental data provided by past researchers. The results produced by GWO algorithm are contrasted to that of previous studies that used other optimization algorithms. The GWO algorithm program is compiled in OCTAVE® and runs on an Intel® Core™ i3-8145U CPU with 7.70 GB RAM. The settings for the algorithm are as follows: population = 500, function evaluations = 100000. It has been found that the GWO algorithm effectively obtains the optimum for all test functions.

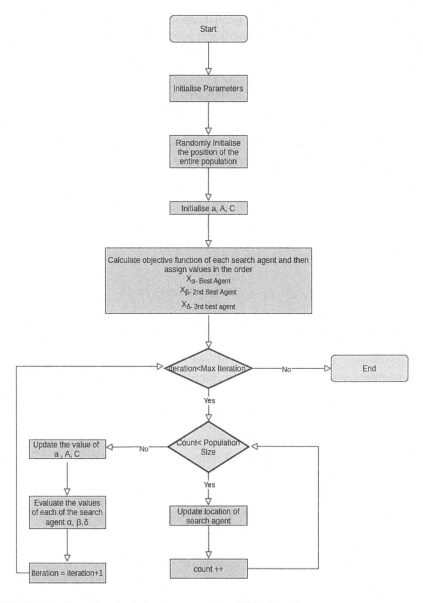

FIGURE 3.2   Flowchart depicting the steps of the GWO algorithm.

3.3.1   CASE: 1

The first case for GWO algorithm optimization of ECM parameters is adopted from
a study on ECM of EN8 steel utilizing an insulated cylindrical solid brass tool with
regulated electrolyte flow and automated tool feed, reported in Reference [11]. In that
experimental study, four ECM process parameters are utilized, and the machining
performance are evaluated using two response parameters, namely MRR and ROC.

TABLE 3.1

Symbols, Units, and Levels of ECM Process Parameters (Case: 1)

			Level				
Parameters	Symbols	Units	−2	−1	0	+1	+2
Electrolyte concentration	A_1	g/l	15	30	45	60	75
Electrolyte flow rate	B_1	l/min	10	11	12	13	14
Voltage	C_1	V	10	15	20	25	30
Inter-electrode gap	D_1	mm	0.4	0.6	0.8	1.0	1.2

Table 3.1 lists the ECM parameters, as well as their levels and values. Mathematical models in actual parameter values are constructed using the experimental results provided in Reference [11] and are presented in Equations (3.6) and (3.7), respectively:

$$
\begin{aligned}
Y_{1\,MRR}\,(g/\min) = {}& 1.19263 + 0.05688\,A_1 - 0.13590\,B_1 + 0.09215\,C_1 - 5.45671\,D_1 \\
& - 0.00004\,A_1^2 + 0.01232\,B_1^2 + 0.00029\,C_1^2 - 0.36444\,D_1^2 \\
& - 0.00365\,A_1 B_1 - 0.00067\,A_1 C_1 + 0.01407\,A_1 D_1 \\
& - 0.01045\,B_1 C_1 + 0.26505\,B_1 D_1 + 0.09247\,C_1 D_1
\end{aligned}
\tag{3.6}
$$

$$
\begin{aligned}
Y_{1\,ROC}\,(mm) = {}& -2.10705 + 0.01065\,A_1 + 0.31849\,B_1 + 0.00266\,C_1 + 0.48742\,D_1 \\
& - 0.00002\,A_1^2 - 0.01223\,B_1^2 + 0.00011\,C_1^2 + 0.08501\,D_1^2 \\
& - 0.00040\,A_1 B_1 - 0.00006\,A_1 C_1 - 0.00199\,A_1 D_1 \\
& + 0.00044\,B_1 C_1 - 0.02656\,B_1 D_1 - 0.00781\,C_1 D_1
\end{aligned}
\tag{3.7}
$$

3.3.1.1 Single-Objective Optimization

Initially, the GWO algorithm is employed for single-objective optimization of MRR and ROC to separately increase MRR and decrease ROC. The GWO algorithm searches for the best process conditions for single-objective optimization of MRR and ROC using Equations (3.6) and (3.7) as objective functions. The maximum MRR and minimum ROC values achieved by GWO algorithm are 1.455 g/min and 0.082 mm, respectively. In Table 3.2, the results are tabulated, along with all the parameter settings and the optimal values. In addition, results obtained by past researchers using different optimization approaches have been summarized in this table. The tabulated data clearly shows that GWO produces a result that is considerably superior to that of the Gauss Jordan Algorithm (GJA) and identical to that of the FPA, TBLO, and ABC algorithms.

The convergence graphs of GWO algorithm for (a) MRR and (b) ROC are presented in Figure 3.3. The GWO algorithm converges to optimal solution after just ten functional evaluations for MRR and ROC optimization. Around 612 functional evaluations converged to the MRR optima, whereas 661 evaluations converged to the ROC optima. The histograms of GWO algorithm functional evaluations for (a) MRR, and (b) ROC are shown in Figure 3.4. The mean MRR and ROC values are 1.114 g/min and 0.1217 mm, with standard deviations of 0.3465 g/min and 0.0352 mm,

TABLE 3.2

Results of Single-Objective Optimization (Case: 1)

Responses	Optimization Goal	Optimization Techniques	Optimal Value	Optimal Parameter Setting			
				A_1	B_1	C_1	D_1
MRR	Maximization	GWO	1.455	75.00	10.00	30.00	1.20
(g/min)		GJA [11]	0.725	57.88	11.98	22.04	1.00
		ABC [15]	1.455	75.00	10.00	30.00	1.20
		TLBO [16]	1.455	75.00	10.00	30.00	1.20
		FPA [31]	1.455	75.00	10.00	30.00	1.20
ROC	Minimization	GWO	0.082	15.00	10.00	10.00	0.40
(mm)		GJA [11]	0.270	17.55	11.05	21.65	0.87
		ABC [15]	0.082	15.00	10.00	10.00	0.40
		TLBO [16]	0.082	15.00	10.00	10.00	0.40
		FPA [31]	0.082	15.00	10.00	10.00	0.40

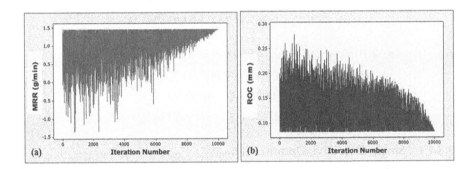

FIGURE 3.3   Convergence graphs of GWO algorithm for (a) MRR and (b) ROC (Case: 1).

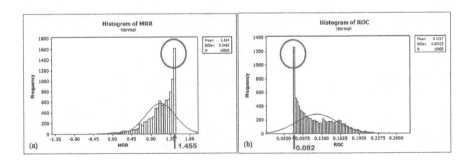

FIGURE 3.4   Histograms of GWO algorithm functional evaluations for (a) MRR and (b) ROC (Case: 1).

respectively, for all 10,000 functional evaluations. The MRR and ROC mean values are comparable to respective optimum values, and the frequency peaks are at their optimal values, ensuring the quality of the convergence. On average, MRR optimization takes 0.73 seconds and ROC optimization takes 0.69 seconds, both of which are less than a second.

3.3.1.2 Multi-Objective Optimization

The multi-objective optimization using the GWO algorithm to simultaneously optimize the MRR and ROC is achieved using the following objective function:

$$minZ = \frac{w_1 \times ROC}{ROC_{min}} - \frac{w_2 \times MRR}{MRR_{max}} \tag{3.8}$$

where w_1 and w_2 are relative significance (so that $w_1 + w_2 = 1$) applied to ROC and MRR, respectively, and the *min* and *max* values in the denominator of Equation (3.8) are the single-objective optimization results of ECM using the GWO algorithm. In this example, w_1 and w_2 are set to 0.5, indicating that MRR and ROC have the same weightage.

The process engineer might pick the weights based on his or her expertise or estimate them using a weighting method such as the analytic hierarchy approach. The multi-objective function Z must be minimized in all situations. The multi-objective function value is reduced to a global minimum of 0.349, and the optimal solutions, as well as all ECM parameters, are furnished in Table 3.3. The GWO algorithm results are compared to those obtained by TBLO, FPA, and ABC algorithms, and identical results are obtained, indicating that further improvement is not feasible within the specified process range. The GWO algorithm convergence graph as well as a histogram of functional evaluations of Z is depicted in Figure 3.5. The response Z is optimized in the 15th functional evolution, and a total of 757 functional evolutions converged to the optima among the 10,000 functional evolutions. The mean value of the response Z obtained is 0.580, with a standard deviation of 0.2106.

3.3.2 CASE: 2

The second case for GWO algorithm optimization of ECM parameters is based on a study on ECM of EN31 steel using an insulated copper tool with axial electrolyte

---

TABLE 3.3

Results of Multi-Objective Optimization (Case: 1)

Conditions	Optimization Techniques	Z_{min}	MRR	ROC	A_1	B_1	C_1	D_1
$w_1 = w_2 = 0.5$	GWO	0.349	0.4408	0.0818	15.00	10.00	10.00	0.40
	TLBO [16]	0.349	0.4408	0.0818	15.00	10.00	10.00	0.40
	FPA [31]	0.349	0.4408	0.0818	15.00	10.00	10.00	0.40
	ABC [15]	0.349	0.4408	0.0818	15.00	10.00	10.00	0.40

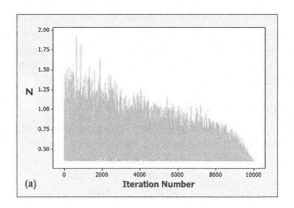

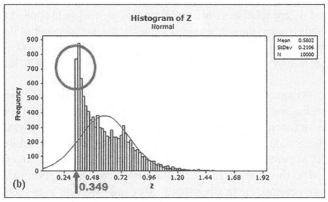

FIGURE 3.5   GWO algorithm (a) convergence graph and (b) histogram of functional evaluations for multi-objective response, Z (Case: 1).

flow through a hole created centrally in the tool, which is published in Reference [17]. In that experimental study, four ECM process parameters are utilized, and the machining performance are evaluated using two response parameters, namely MRR and Ra. Table 3.4 lists the ECM parameters, as well as their levels and values. The mathematical models in actual parameter values are constructed using the experimental

TABLE 3.4

Symbols, Units, and Levels of ECM Process Parameters (Case: 2)

			Level		
Parameters	**Units**	**Symbols**	**−1**	**0**	**+1**
Electrolyte concentration	%	E_I	15	20	25
Voltage	V	F_I	8	10	12
Feed rate	mm/min	G_I	0.10	0.20	0.30
Inter-electrode gap	mm	H_I	0.20	0.25	0.30

results provided in Reference [17] and are presented in Equations (3.9) and (3.10), respectively:

$$Y_{2\,MRR}\,(g/\min) = 3.0085 - 0.4624\,E_1 + 0.1695\,F_1 - 2.6797\,G_1 + 8.2388\,H_1$$
$$+ 0.0119\,E_1^2 - 0.0114\,F_1^2 + 10.9425\,G_1^2 - 19.1300\,H_1^2$$
$$+ 0.0007\,E_1 F_1 + 0.0535\,E_1 G_1 - 0.1273\,E_1 H_1$$
$$+ 0.0081\,F_1 G_1 + 0.2856\,F_1 H_1 - 8.6460\,G_1 H_1 \tag{3.9}$$

$$Y_{2\,Ra}\,(\mu m) = 11.8339 - 0.3788\,E_1 + 0.1375\,F_1 - 2.1900\,G_1 - 51.7693\,H_1$$
$$- 0.0011\,E_1^2 - 0.0130\,F_1^2 - 5.1382\,G_1^2 + 37.9873\,H_1^2$$
$$+ 0.0057\,E_1 F_1 + 0.0787\,E_1 G_1 + 1.3650\,E_1 H_1$$
$$- 0.0147\,F_1 G_1 + 0.2918\,F_1 H_1 + 8.6463\,G_1 H_1 \tag{3.10}$$

3.3.2.1   Single-Objective Optimization

First, the GWO algorithm is used for single-objective optimization of MRR and Ra to separately increase MRR and decrease Ra. The GWO algorithm determines for the optimal parameters for single-objective optimization of MRR and Ra using Equations (3.9) and (3.10) as objective functions. The maximum MRR and minimum Ra achieved by the GWO algorithm are 1.089 g/min and 1.496 μm, respectively. Table 3.5 furnishes the optimal results as well as the ECM parameter values. Furthermore, the results of previous studies utilizing various optimization techniques, such as the FPA and ABC algorithms, have been presented in this table. It is witnessed that they anticipate the same response values and the same level of process parameters at the optimal process condition.

Figure 3.6 depicts the GWO algorithm convergence graphs for (a) MRR and (b) Ra. The GWO algorithm is observed to converge to the optimal solution after 8,587 functional evaluations for MRR optimization and just 20 functional evaluations for Ra optimization. Around 54 functional evaluations converged to the MRR optima, whereas 597 evaluations converged to the Ra optima. The histograms of GWO algorithm functional evaluations for (a) MRR, and (b) Ra are shown in Figure 3.7.

TABLE 3.5

Results of Single-Objective Optimization (Case: 2)

Responses	Optimization Goal	Optimization Techniques	Optimal Value	Optimal Parameter Setting			
				E_1	F_1	G_1	H_1
MRR	Maximization	GWO	1.089	25.00	10.99	0.30	0.21
(g/min)		FPA [18]	1.089	25.00	10.99	0.30	0.21
		ABC [17]	1.086	25.00	11.00	0.30	0.20
Ra	Minimization	GWO	1.496	25.00	8.00	0.30	0.20
(μm)		FPA [18]	1.496	25.00	8.00	0.30	0.20
		ABC [17]	1.490	25.00	8.00	0.30	0.20

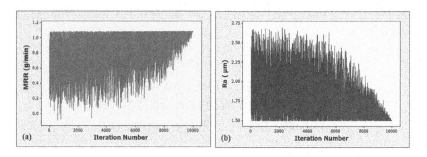

FIGURE 3.6 Convergence graphs of GWO algorithm for (a) MRR, and (b) Ra (Case: 2).

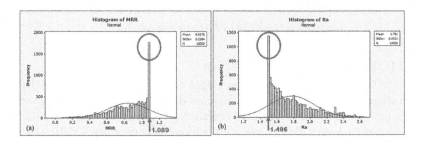

FIGURE 3.7 Histograms of GWO algorithm functional evaluations for (a) MRR, and (b) Ra (Case: 2).

The mean MRR and Ra values are 0.837 g/min and 1.781 μm, respectively, for all 10,000 functional evaluations. On average, MRR optimization takes 0.93 seconds and Ra optimization takes 0.91 seconds, both of which are less than a second.

3.3.2.2 Multi-Objective Optimization

The multi-objective optimization using the GWO algorithm to simultaneously optimize the MRR and Ra is achieved using the following objective function:

$$minZ = \frac{w_1 \times ROC}{ROC_{min}} - \frac{w_2 \times MRR}{MRR_{max}} \tag{3.11}$$

where w_1 and w_2 are the weights applied to Ra and MRR, respectively. The values of w_1 and w_2 are set to 0.5, indicating that MRR and Ra have the same weightage. In all cases, the multi-objective function Z must be minimized. The multi-objective function value reaches to a global minimum of 0.043, and the optimal solutions, as well as all ECM parameters, are furnished in Table 3.6. The GWO algorithm results are compared to those obtained by FPA, and ABC algorithms and identical results are obtained. Figure 3.8 depicts the GWO algorithm convergence diagram as well as a histogram of functional evaluations of response Z. The mean value of the response Z obtained is 0.233, with a standard deviation of 0.166. Even though the mean value of optimized results deviates significantly from the optimum results, the frequency peak remains at the global minimum of 0.043.

TABLE 3.6

Results of Multi-Objective Optimization (Case: 2)

Condition	Optimization Techniques	Z_{min}	MRR	Ra	E_1	F_1	G_1	H_1
$w_1 = w_2 = 0.5$	GWO	0.0436	0.994	1.496	25.00	8.00	0.30	0.20
	FPA [18]	0.0436	0.994	1.496	25.00	8.00	0.30	0.20
	ABC [17]	0.0436	0.994	1.490	25.00	8.00	0.30	0.20

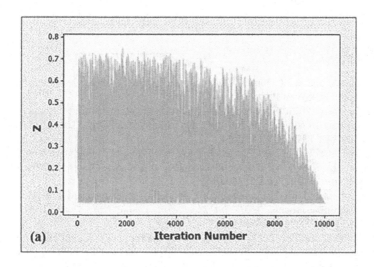

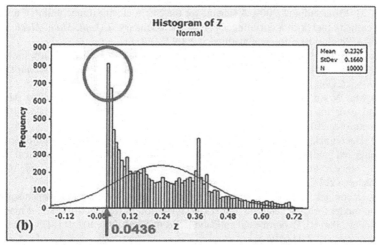

FIGURE 3.8 GWO algorithm (a) convergence graph and (b) histogram of functional evaluations for multi-objective response, Z (Case: 2).

3.4 CONCLUSION

In this work, the GWO algorithm is employed to optimize ECM process parameters on a single and multi-objective basis. Different ECM response characteristics are set in two independent case studies in relation to machining parameters such as voltage, inter-electrode gap, feed rate, electrolyte concentration, and electrolyte flow rate. In the first case, the MRR is maximized, and Ra is minimized simultaneously, while in the second case, the MRR is maximized, and ROC is minimized simultaneously. The case studies on ECM process optimization demonstrate the GWO algorithm's ability to precisely predict the optimal sets of ECM parameters to enhance machining performance and quality. The GWO algorithm results are compared to those obtained by GJA, ABC, FPA, and TLBO algorithms, demonstrating the efficacy of GWO for optimizing ECM parameters to improve performance characteristics.

REFERENCES

[1] A. Ghosh, and A. K. Mallik. 2016. *Manufacturing Science*, second ed. New Delhi: Affiliated East-West Press.
[2] Z. Pandilov. 2018. Application of electrochemical machining for materials used in extreme conditions. *IOP Conf. Ser. Mater. Sci. Eng.*, 329 (1), 012014. doi: 10.1088/1757-899X/329/1/012014.
[3] Z. Xu, and Y. Wang. 2019. Electrochemical machining of complex components of aero-engines: Developments, trends, and technological advances. *Chinese J. Aeronaut.*, 34 (2), 28–53. doi: 10.1016/j.cja.2019.09.016.
[4] R. Tsuboi, and M. Yamamoto. 2010. Modeling and applications of electrochemical machining process. *ASME Int. Mech. Eng. Congr. Expo. Proc.*, 4, 377–384. doi: 10.1115/IMECE2009-12552.
[5] N. M. Minazetdinov. 2009. A hydrodynamic interpretation of a problem in the theory of the dimensional electrochemical machining of metals. *J. Appl. Math. Mech.*, 73 (1), 41–47. doi: 10.1016/j.jappmathmech.2009.03.009.
[6] N. M. Minazetdinov. 2009. A scheme for the electrochemical machining of metals by a cathode tool with a curvilinear part of the boundary. *J. Appl. Math. Mech.*, 73 (5), 592–598. doi: 10.1016/j.jappmathmech.2009.11.012.
[7] Z. Li, and Z. Niu. 2007. Convergence analysis of the numerical solution for cathode design of aero-engine blades in electrochemical machining. *Chinese J. Aeronaut.*, 20 (6), 570–576. doi: 10.1016/S1000-9361(07)60084-3.
[8] N. Ma, W. Xu, X. Wang, and B. Tao. 2010. Pulse electrochemical finishing: Modeling and experiment. *J. Mater. Process. Technol.*, 210 (6–7), 852–857. doi: 10.1016/j.jmatprotec.2010.01.016.
[9] D. Deconinck, S. Van Damme, and J. Deconinck. 2012. A temperature dependent multi-ion model for time accurate numerical simulation of the electrochemical machining process. Part I: Theoretical basis. *Electrochim. Acta*, 60, 321–328. doi: 10.1016/j.electacta.2011.11.070.
[10] D. Deconinck, W. Hoogsteen, and J. Deconinck. 2013. A temperature dependent multi-ion model for time accurate numerical simulation of the electrochemical machining process. Part III: Experimental validation. *Electrochim. Acta*, 103, 161–173. doi: 10.1016/j.electacta.2013.04.059.
[11] B. Bhattacharyya, and S. K. Sorkhel. 1998. Investigation for controlled electrochemical machining through response surface methodology-based approach. *J. Mater. Process. Technol.*, 86 (1–3), 200–207. doi: 10.1016/S0924-0136(98)00311-2.

[12] B. Bhattacharyya, M. Malapati, J. Munda, and A. Sarkar. 2007. Influence of tool vibration on machining performance in electrochemical micro-machining of copper. *Int. J. Mach. Tools Manuf.*, 47 (2), 335–342. doi: 10.1016/j.ijmachtools.2006.03.005.

[13] R. V. Rao, P. J. Pawar, and R. Shankar. 2008. Multi-objective optimization of electrochemical machining process parameters using a particle swarm optimization algorithm. *Proc. Inst. Mech. Eng. Part B J. Eng. Manuf.*, 222 (8), 949–958. doi: 10.1243/09544054JEM1158.

[14] P. Asokan, R. R. Kumar, R. Jeyapaul, and M. Santhi. 2008. Development of multi-objective optimization models for electrochemical machining process. *Int. J. Adv. Manuf. Technol.*, 39 (1–2), 55–63. doi: 10.1007/s00170-007-1204-8.

[15] S. Samanta, and S. Chakraborty. 2011. Parametric optimization of some non-traditional machining processes using artificial bee colony algorithm. *Eng. Appl. Artif. Intell.*, 24 (6), 946–957. doi: 10.1016/j.engappai.2011.03.009.

[16] R. V. Rao, and V.D. Kalyankar. 2011. Parameters optimization of advanced machining processes using TLBO algorithm. *EPMM Singapore*, 21–32. doi: 10.32738/ceppm. 201109.0003.

[17] M. K. Das, K. Kumar, T. K. Barman, and P. Sahoo. 2014. Investigation on electrochemical machining of EN31 steel for optimization of MRR and surface roughness using artificial bee colony algorithm. *Procedia Eng.*, 97, 1587–1596. doi: 10.1016/ j.proeng.2014.12.309.

[18] B. Acherjee, D. Maity, and A. S. Kuar. 2020. Optimization of correlated and conflicting responses of ECM process using flower pollination algorithm. *Int. J. Appl. Metaheuristic Comput.*, 11 (4), 1–15. doi: 10.4018/IJAMC.2020100101.

[19] T. Sekar, V. Sathiyamoorthy, K. Muthusamy, A. Sivakumar, and S. Balamurugan. 2021. Artificial fish swarm algorithm driven optimization for copper-nano particles suspended sodium nitrate electrolyte enabled ECM on die tool steel. In: *Futuristic Trends in Intelligent Manufacturing*, edited by K. Palanikumar, E. Natarajan, R. Sengottuvelu, and J. P. Davim, 47–60. Cham: Springer. doi: 10.1007/978-3-030-70009-6_5.

[20] H. Faris, I. Aljarah, M. A. Al-Betar, and S. Mirjalili. 2018. Grey wolf optimizer: A review of recent variants and applications. *Neural Comput. Appl.*, 30 (2), 413–435. doi: 10.1007/s00521-017-3272-5.

[21] S. X. Li, and J. S. Wang. 2015. Dynamic modeling of steam condenser and design of PI controller based on grey wolf optimizer. *Math. Probl. Eng.*, 2015, 120975. doi: 10.1155/2015/120975.

[22] S. Yadav, S. K. Verma, and S. K. Nagar. 2016. Optimized PID controller for magnetic levitation system. *IFAC-PapersOnLine*, 49 (1), 778–782. doi: 10.1016/j.ifacol.2016.03.151.

[23] K. R. Das, D. Das, and J. Das. 2016. Optimal tuning of PID controller using GWO algorithm for speed control in DC motor. *Int. Conf. Soft Comput. Tech. Implementations, ICSCTI 2015*, 108–112. doi: 10.1109/ICSCTI.2015.7489575.

[24] A. Kumar, S. Pant, and M. Ram. 2017. System reliability optimization using gray wolf optimizer algorithm. *Qual. Reliab. Eng. Int.*, 33 (7), 1327–1335. doi: 10.1002/qre.2107.

[25] S. Khalilpourazari, and S. Khalilpourazary. 2018. Optimization of production time in the multi-pass milling process via a robust grey wolf optimizer. *Neural Comput. Appl.*, 29 (12), 1321–1336. doi: 10.1007/s00521-016-2644-6.

[26] G. M. Komaki, and V. Kayvanfar. 2015. Grey wolf optimizer algorithm for the two-stage assembly flow shop scheduling problem with release time. *J. Comput. Sci.*, 8, 109–120. doi: 10.1016/j.jocs.2015.03.011.

[27] J. Rameshkumar, S. Ganesan, S. Subramanian, and M. Abirami. 2015. Short-term unit consignment solution using real-coded grey wolf algorithm. *Aust. J. Electr. Electron. Eng.*, 13 (1), 54–66. doi: 10.1080/1448837X.2015.1092933.

[28] S. Mirjalili, S. M. Mirjalili, and A. Lewis. 2014. Grey wolf optimizer. *Adv. Eng. Softw.*, 69, 46–61. doi: 10.1016/j.advengsoft.2013.12.007.

[29] J. M. Packard. 2003. Wolf behavior: Reproductive, social, and intelligent. In: *Wolves: Behavior, Ecology and Conservation*, edited by L. David Mech and Luigi Boitani, 35–65. Chicago: University of Chicago Press. doi: 10.7208/9780226516981-006.

[30] C. Muro, R. Escobedo, L. Spector, and R. P. Coppinger. 2011. Wolf-pack (Canis lupus) hunting strategies emerge from simple rules in computational simulations. *Behav. Processes*, 88 (3), 192–197. doi: 10.1016/j.beproc.2011.09.006.

[31] B. Acherjee, D. Maity, A. S. Kuar, and M. K. Datta. 2019. Application of flower pollination algorithm for optimization of ECM process parameters. In: *Optimization for Engineering Problems*, edited by K. Kumar and J. P. Davim, 17–37. ISTE – Wiley. doi: 10.1002/9781119644552.ch2.

4 Simulation for Decision-Making in the Orthogonal Turning Operation of Three Different Aluminum Alloys Using John-Cook Plasticity Model and Damage Laws

Padmanabhan K, Saraschandra M,
Prudvi Krishna M, Bhargav Avinash G
VIT University, Vellore, Tamil Nadu, India

CONTENTS

DOI: 10.1201/9781003121640-6

4.1   INTRODUCTION

Machining is the process done for the removal of the excess material from the casted or formed components to meet the required specifications. During machining, the temperature at the tool–chip interface, in particular, will be very high due to adiabatic heat generation. Temperature rise, heat partition and transferred heat at the contact between tool and chip are key parameters for a precise prediction of tool wear, tool life and surface integrity. The heat generation mainly depends on the machinability of the workpiece and the thermo-physical properties of cutting material which are important factors for predicting the temperature distribution and heat dissipation at the tool–workpiece interface. Physical phenomena occurring at this interface depend on local conditions of stress (contact pressure and frictional stress): sliding velocity, cutting temperature and local properties of the tool–work piece. The complexity of the thermomechanical processes occurring during the chip-formation process makes it difficult to estimate the heat exchange at the cutting zone. There has been a lot of growth in the field of automotive and aerospace industries in the last three decades, from the high strength to weight ratio materials like the aluminum alloys that show widespread applications. They offer a number of different mechanical and thermal properties. In addition, they are relatively easy-to-shape materials, especially in material removal processes, such as machining aluminum alloys as this class is considered as a family of materials offering the highest levels of machinability. Innovative research is being pursued for the optimized working parameters in dry machining of aluminum alloys for an eco-friendly approach. Different theories have been proposed for aluminum alloys machining. Eulerian approach,

Lagrangian approach and Arbitrary Lagrangian Eulerian (ALE) approach are the most widely used and promising approaches in the finite element method-based simulations, when a nonlinear/explicit case is encountered as in turning.

NOMENCLATURE

σ_y	= Equivalent Von-Mises flow stress	f	= Feed rate
$[\bar{\varepsilon}^p]$	= Equivalent plastic strain	t	= Cutting time
n	= Strain hardening index	d	= Depth of cut
$\dot{\bar{\varepsilon}}^p$	= Equivalent plastic strain rate	m	= An index determined
$\dot{\varepsilon}^0$	= Initial dimensionless plastic strain rate		from $\dot{\bar{\varepsilon}}^p$ and T
T_{melt}	= Melt temperature		
T_{room}	= Workpiece transition temperature		

Markopoulous (2012) in his "Finite Element Methods in Machining Process" included different types of material models, friction models for machining, adaptive machining and applications of FEM in metal cutting. Aluminum and its alloys have been discussed in detail for their machinability in general and turning in particular in an ASM metals handbook (Krenzer, 1989). Frittz Klocke (2011) included different finite element models for continuous, segmented and discontinuous chip formations. Simulation studies of turning of aluminum cast alloys using PCD tools were carried out and discussed in Reference (Frager da Silva and Reis, 2017). Mabrouki (2008) worked on the AA-2024 aluminum alloy and how the finite element method could be considered for uniting the damage and plastic-elastic fracture. A numerical model confirming the increase in the partition of the chip with the increase in the machining parameters and chip shape in the overall machining process when the fragmentation of the chip takes place was investigated.

Evdokimov (2014) worked on computation of temperature measures in the tool and with the changing parameters using ANSYS and restrained stresses within the work using Abaqus. Panatale et al. (2004) gave a wide-cut information of how to model the orthogonal machining in which the chipping of the work is purely based on the inherent nature of the work and mathematical model stresses on the above and the entire description of what machinability is a complete procedure for the simulation of the cutting operation. Starting from the identification of the constitutive and damage laws of the material, a numerical model is built, for which it must be emphasized that the formation of the chip involves the intrinsic behavior of the material, then bringing a comprehensive model of what is called "machinability."

Sawarkar and Boob (2004) have explicated how the machining parameters and aesthetics will reduce the stresses within the work and the time consumption can be optimized with a better mathematical computation using FEM and Abaqus. Xiao et al. (2011) simulated the metal cutting and chip formation of flat-groove specimen and effect of stress tri-axiality factors on strain rate and strain energy density using the Johnson-Cook model. Bhoyar and Kamble (2013) worked on FEM of turning process to obtain the effect of cutting speed, feed rate and tool geometry, and cutting forces and specific cutting energy were evaluated and temperature was obtained from tool–workpiece contact region.

Klocke and Puls (2013) worked on turning of workpieces using C45E and AISI 1045 cutting tools. FEM was done using DEFORM™ to evaluate the thermal deformation on workpiece, and these values were verified using the thermal imaging method. The effect of minimum quantity lubrication under diffferent parameters in the turning of AA-7075 and AA-2024 aluminum alloys was investigated by Cakir et al. (2016).

Shrot and Baker (2012) investigated the determination of Johnson-Cook parameters from machining simulations. The effect of rake angle between the workpiece and the tool on Johnson-Cook material constants and their impact on the cutting process parameters of aluminum 2024-T3 alloy machining simulation was simulated and analyzed by Daoud et al. (2015). Ijaz et al. (2017) have investigated the modified Johnson-Cook plasticity model with damage evolution for its application to turning and orthogonal turning simulation of 2XXX aluminum alloys. Asad et al. (2014) have conducted research on the turning modeling and simulation of an aerospace grade aluminum alloy using 2D and 3D finite element methods for orthogonal turning. As the current investigation deals with the 2D and 3D aspects of turning, these references and their support are considered important.

In this investigation, with the help of the Abaqus® software, the thermomechanical analysis for orthogonal turning of aluminum alloys was conducted. The resulting heat generation and its prediction form the main interest of this chapter where the cutting forces are also evaluated. In case lower order elements are used, which an explicit analysis always prefers, the mass matrix is also a lumped matrix, or a diagonal matrix, whose inversion is a single step process of just making the diagonal elements reciprocal. This is very easily done but the disadvantage is that Euler time integration scheme is not used in this approach generally because it is not unconditionally stable. So, small steps are normally taken for the explicit analysis to understand the heat generation during turning. The methodology adopted scope of the analysis and outcomes are discussed in the following sections.

4.2 METHODOLOGY

The following methods are generally used for meshing in FEM packages.

4.2.1 Lagrangian Method

In Lagrangian method, the finite element mesh and workpiece elements are united and perhaps continuums are large. It is widely used for its fast computations and no transportation of the work with mesh is to be deliberated. The material constraints need not be preset.

4.2.2   EULERIAN METHOD

The material moves by the given mesh. It is necessary to calculate the work parameters at the required locations. Grid deformations are formed and the time step for solution will be high and the chip formation is determined to the feed conditions prior to simulation. It is often used in hydrodynamic problems.

4.2.3   ARBITRARY LAGRANGIAN EULERIAN (ALE) METHOD

Arbitrary Lagrangian Eulerian method is a synergy of both Eulerian (used for mocking up the surrounding area of the tip of the tool) and Lagrangian methods (used for designing the free flow of the material at its limits). It is exclusively used in shell elements for explicit dynamics.

4.2.4   SMOOTH PARTICLE HYDRODYNAMICS METHOD (SPHM)

It is a pure Lagrangian method with no meshing and grid. It is easy to compute large deformations because of a lack of mesh and calculation of interactions between separated material particles.

Raja and Baskar (2011) have interestingly carried out a different study on particle swarm optimization technique for determinining the optimal machining parameters of different workpiece materials in turning operation.

The analysis in this investigation is being carried out using the Lagrangian method because ALE is suitable only for shell elements and the Eulerian and Smooth particle hydrodynamics are not quite suitable for structural analysis involving thermomechanical inputs.

4.2.4.1   Element Modelling

In Abaqus, different types of meshing methods and elements are used. For meshing, solid model hexahedral brick elements and tetrahedral brick elements are available in the software. Here, hexahedral brick is used to mesh both the workpiece and tool. In the metal-cutting operation, the work, chip and tool are major parts and these are meshed with the element type which is an explicit couple field element. The tool is meshed with the standard couple field element and remaining workpiece is meshed with standard 3D-stress element.

4.2.4.2   Material Modelling

The workpiece undergoes deformation at every element the tool passes by, according to the tool preset movements, following a given explicit algorithm. Johnson-Cook turns out to be the best method for the simulation of machining, to study problems on fast deformations, stresses and strains due to adiabatic heat rise (i.e. large strain rates, criteria of equivalent strain).

Equations and constants that Johnson-Cook derived for strain rate in the machining are published in previous papers by noteworthy researchers (Shrot and Baker, 2012; Asad et al., 2014; Ijaz et al., 2017; Camilo et al., 2019).

$$\sigma_y = \left[A + B * \left(\overline{\varepsilon}^{pn} \right) \right] * \left[1 + c * \frac{\dot{\overline{\varepsilon}}^p}{\dot{\varepsilon}^0} \right] * \left[1 - \left(\frac{T - T_{room}}{T_{melt} - T_{room}} \right) * m \right] \qquad (4.1)$$

$$\varepsilon_{failure} = \left[D_1 + D_2 * e^{D_3 * \sigma^*} \right] * \left[1 + D_4 * \ln(\ddot{e}) \right] * \left[1 + D_5 * \left(\frac{T - T_{room}}{T_{melt} - T_{room}} \right) \right] \qquad (4.2)$$

where

σ_y = Equivalent Von-Mises flow stress in MPa
$\varepsilon_{failure}$ = Strain at fracture
$[\overline{\varepsilon}^p]$ = Equivalent plastic strain
n = Strain hardening index
$\dot{\overline{\varepsilon}}^p$ = Equivalent plastic strain rate in s^{-1}
$\dot{\varepsilon}^0$ = Initial dimensionless plastic strain rate
ë = Ratio of equivalent plastic strain rate and initial dimensionless plastic strain rate
σ^* = Ratio of pressure to effective stress
T_{melt} = Melt temperature in degree Celsius
T_{room} = Workpiece transition temperature in degree Celsius.
m = An index or material constant determined from known temperatures and strain rates.

$A, B, C, D_1, D_2, D_3, D_4, D_5$ are constants. Here for a time-step and path-dependent damage analysis, the constant C is not required to be used.

Here AA-2024, AA-6061, AA-7075 are alloy materials assigned to the workpiece and SS-4340 is the nickel-chromium-molybdenum stainless steel material assigned to tool. AA-2024 contains Cu 4.4, Mn 0.6 and Mg 1.5, while AA-6061 contains Si 0.6, Mg 1.0 and Cu 0.28 and AA-7075 contains Zn 5.6, Mg 2.5, Cu 1.6 and Cr 0.23 all by weight percentage. The main remaining constituent is aluminum in all these alloys. Turning is considered for their untreated and as cast conditions here.

Tables 4.1–4.4 indicate the material properties obtained from the ASM website (ASM Vol. 16, 1989) and various publications for the three alloys AA-2024, AA-6061 and AA-7075 (Shrot and Baker, 2012; Asad et al., 2014; Ijaz et al., 2017; Camilo et al., 2019).

It is to be noted that though some of the parameters are variables with respect to temperature, such as the thermal expansion coefficient, the available data is quite limited in this regard.

TABLE 4.1
Material Properties of Al Alloys

Material	Density (kg/m³)	Young's Modulus GPa	Poisson's Ratio	Specific Heat (J/kg.C)	Thermal Conductivity (W/m.C)	Coefficient of Thermal Expansion (at 300°C) 10⁻⁶/c
AA-2024	2700	73	0.34	881	164	14
AA-6061	2780	70	0.33	942	154	35
AA-7075	2810	71	0.32	858	120	22

TABLE 4.2
Johnson-Cook Constants of Al Alloys

Material	A(MPa)	B(MPa)	N	M	T_{melt} (°C)
AA-2024	352	440	0.42	1	520
AA-6061	324	114	0.42	1.34	655
AA-7075	527	575	0.72	1.61	621

TABLE 4.3
Johnson-Cook Damage Equation Parameters

Material	D_1	D_2	D_3	D_4	D_5
AA-2024	0.13	0.13	−1.5	0.011	0
AA-6061	−0.77	1.45	−0.47	0	0
AA-7075	0.096	0.049	3.465	0.016	1.099

TABLE 4.4
Properties of Cutting Tool Material

Material	Density (kg/m³)	Young's Modulus (GPa)	Poisson's ratio	Specific Heat (J/kg°C)	Thermal Conductivity (W/m.°C)
Steel-4340	7870	200	0.29	475	44.5

4.2.4.3   Meshing and Boundary Conditions

For the finite element analysis of metal cutting using Abaqus dynamic, temp, explicit process is used. In this analysis, contact pair is defined between the chip and the tool by node-to-surface definition with friction followed by Coulomb's law, heat generation, and thermal conductance. After defining the contact pair, time step of

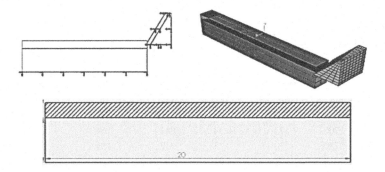

FIGURE 4.1   Meshing model of workpiece and tool with boundary conditions for orthogonal turning.

0.001s is defined. Boundary conditions are applied on the bottom of a workpiece as encastre—that is, fixed like a beam on both the ends, tool as fixed and in the final step, the tool is said to move with a distance equal to the length of the workpiece defined here in the circumferential direction. A small circumferential hoop section distance of a long cylindrical workpiece was assumed to be planar (and not curved over an infinitesimal surface area) and the circumferential distance was orthogonally turned with a tool of the described geometry. The tool path in the region along the surface is described in Figure 4.1. The 2D and 3D drawings and plots of Figure 4.1 have been captioned. The simulation package used is Abaqus and a nonlinear simulation was carried out.

4.2.4.4   Assumptions
- The following are the reasons for carrying out 3D simulations: (a) In 2D simulations, the applicability to machining would be limited and one cannot provide surface-to-surface contact nonlinearity for analyzing tools and chip surface morphology or profiling; (b) Shapes of the chips or turnings obtained and the temperature profile of the tool and the chip or turnings due to frictional heating cannot be best evaluated in 2D plots. However, for positional and cross-sectional details and visualization, some of the images have been retained in 2D views.
- The dynamic coefficient of friction between steel and all three aluminum alloys is taken as 0.46. This value is initially expected to be high as assumed and decreases with the same tool turning the workpiece. Asad et al. (2017) have arrived at an initial coefficient of friction for the orthogonal turning modeling and simulation of an aerospace-grade aluminum alloy using 2D and 3D finite element method. In this investigation, an initial value of 0.46 has been chosen based on the initial conditions that always prevail during the turning of the aluminum workpiece surface with an R_{max} of 6.57 microns and an R_a of 2.54 microns, keeping in mind that the tool path is one way and the dynamic for the machining parameters is shown next.

- Workpiece was not rotated and is defined as prismatic in shape. Machining parameters were

$$\text{Feed } (f) = 20 \text{ m/s}$$

$$\text{Cutting time } (t) = 1 \text{ milli second}$$

$$\text{Depth of cut } (d) = 1 \text{ mm}$$

$$\text{Metal Removal Rate} = [\text{Area of Work piece} - \text{Area of Cutting Portion}] * \text{Feed}$$

$$= \left[8\text{mm}^2 - 2\text{mm}^2 \right] * 20 * 10^3 \text{mm/s}$$

$$= 12 * 10^4 \text{mm}^3 /s$$

Explicit dynamic analysis is adapted here; the part above the workpiece is only meshed with explicit elements. Heat generation by friction is assumed. The thermal heat generation is assumed to be adiabatic under the present conditions where each millisecond events are recorded. The high plastic strain rate is a manifestation of the adiabatic heat rise due to turning that causes ductility. High feed and a depth of cut on the higher side were assumed for the simulation that is within the upper limit and is normally used for aluminum turning. The heat generation and cutting forces are expected to be on the higher side compared to some of the simulations that have not used the Johnson-Cook model for the temperature and stresses observed. Not withstanding the abovesaid argument, some investigators have optimized the process parameters in the computer numerical control (CNC) turning of aluminum alloys using the hybrid RSM-cum-TLBO approach and also probed into a multiobjective optimization of cutting conditions when turning aluinium alloys 1350-O and 7075-T6 grades using genetic algorithm (Santos et al., 2015; Rudrapati et al., 2016). The two investigations cited here have optimized the turning operation, specifically for an alloy mentioned and not for the alloys under investigation that undergo orthogonal turning, which is known to produce higher temperatures and cutting forces due to the nature of tool contact with workpiece.

4.3 RESULTS AND DISCUSSION

The results and discussion from our investigations provide some salient details on the temperature profiles, cutting forces, Von-Mises and shear stresses and possible influences of discontinuities near the melting point, such as a change in the assumed thermal properties and phases. Javidika et al. (2020) report on the impact of tool geometry and cutting conditions in the turning of aluminum alloys 6061-T6. This interesting study is an experimentally validated numerical study that lends support to our chosen machining parameters. The influence of the cutting speed on the temperature and cutting forces at the precision turning of non-ferrous metals using cutters with round diamond hard-alloyed plates have been studied extensively by Stakhniv and colleagues (Stakhniv and Devin, 2019). The machining parameters are not comparable in the precise sense, however, which gives them a lower temperature at the frictional interface. In our investigation, as the orthogonal turning parameters were on the higher side, especially the feed and the depth of cut, the assumed Johnson-Cook parameters are suspected to go through a noticeable change due to discontinuities in the heat capacity and thermal conductivity as the melting temperatures are approached, governed by the laws of physics. Here, the rigid visco-plastic behavior breaks down once softening sets in. However, in simulations done based on assumed Johnson-Cook constants, the effects due to discontinuities near the melting temperatures caused by a higher feed, cutting force, speed and/or depth of cut are not accounted for, which gives us a higher range of simulated temperatures due to machining than the experimentally determined values. Discontinuities near the melting point in thermal properties generally decrease the maximum observed temperature. Kumar et al. (2019) have documented the influence of cutting parameters on cutting forces and surface roughness in dry turning of aluminum alloys using PCD and different coated steels. In the subsequent sections, our own derivation of the temperature profile and cutting forces are discussed.

4.3.1 FEM Plots of AA-2024 Alloy

4.3.1.1 Von-Mises Plot

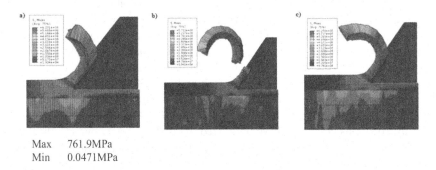

Max   761.9MPa
Min   0.0471MPa

FIGURE 4.2 Von-Mises stresses at (a) 0.25 ms, (b) 0.5 ms and (c) 0.75 ms.

4.3.1.2 Temperature Plot

Max 1357^0C (Tool)
Min 392.7^0C

FIGURE 4.3 Temperature plots at (a) 0.25 ms, (b) 0.5 ms and (c) 0.75 ms shear stress plot.

4.3.1.3 Shear Strain Plot

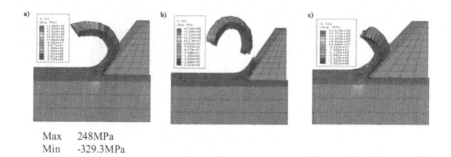

Max   248MPa
Min   -329.3MPa

FIGURE 4.4   Shear stress at (a) 0.25 ms, (b) 0.5 ms and (c) 0.75 ms.

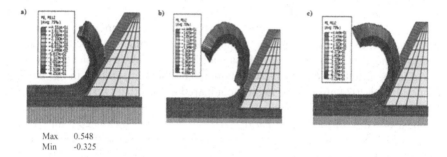

Max   0.548
Min   -0.325

FIGURE 4.5   Shear strain at (a) 0.25 ms, (b) 0.5 ms and (c) 0.75 ms.

In AA-2024, the chip formation is generally observed in a continuous form, although at a time frame of 0.5 ms, the chip is cut in a serrated and discontinuous form. Depth of cut is less and longer chips are produced at a 45° rake angle and 10° cutting-edge angle. Maximum von-Mises stress is observed as 761.9 MPa at the tool–chip interface at 0.95 ms time after turning. Figures 4.2–4.5 depict the temperature, shear stress and shear strain parameters derived from the finite element analysis.

4.3.1.4 Temperature on Cutting Tool

From Figure 4.6, it is seen that the maximum temperature observed on select local locations on the cutting tool is 1357°C, whereas the melting point of the molybdenum-based steel 4340 is closer to 1493°C. Tool wear commences quite below the melting point due to grain coarsening and plasticity. The maximum shear stress is observed at the tool–chip interface. Experimental measurement of tempeature in the orthogonal turning of $AlCu_3MgMnPb$ aluminum alloys has been reported by Dubovska and colleague (2015). Though the composition of the alloy is slightly different from ours, the first addition, viz. copper content is similar. They have reported a tool tip temperature of up to 1000°C measured experimentally. Though there are many methods to measure the temperature of a tool or the interfeace,

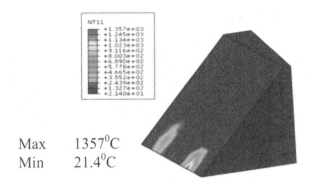

Max 1357^0C
Min 21.4^0C

FIGURE 4.6   Temperature distribution on cutting tool after machining.

Dubovska and Majerik (2015) have used the FLIR (Forward Looking Infrared) camera, which is suitably positioned to detect the interface and tool temperature radiatively. This is bound to be lower than the simulated temperature (as per our investigation) as only the radiative component is experimentally detected due to frictional heating. Besides, the FLIR camcorders use a lower wavelength IR compared to the usual thermographic techniques. The tool temperature in our simulation of the Al-2024 alloy is the highest compared to the other two alloys described later, due to highly conductive additives present in the workpiece that improve the heat transfer with the tool. Paszta (2017) reports the workpiece temperatures achieved for AL-2017 and Al-7075 alloys during turning using the Johnson-Cook model in a DEFORM 3D software. Workpiece temperatures of upto 550°C were reached. The same techniques are expected to detect a higher temperature in orthogonal turning of the same alloys.

4.3.1.5   Von-Mises Stress and Chip Formation

Figure 4.7 and the plot represent the surface soon after machining and how the temperature gradient is manifested in machining. It is observed that so many regions

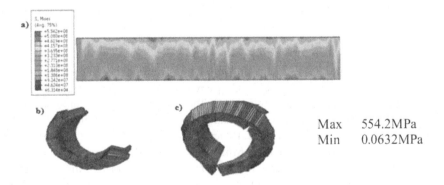

Max   554.2MPa
Min   0.0632MPa

FIGURE 4.7   (a) Von-Mises stress distribution on surface of workpiece after machining, (b) chip formed at 0.5 ms and (c) chip formed at 1 ms.

occur where high Von-Mises stresses are induced. These above figures represent the chips formed during machining. A surface-to-surface contact assumption in 3D FEM is required to model and simulate the chip formation and study its turnings and profile. We find that a similar study has been made by Asad et al. (2014). First chip is formed at 0.5 ms and the second chip is formed at 1 ms. Chip discontinuity is observed in the second chip formed at 1 ms. This is normally caused due to discontinuities in the thermal conductivity and diffusivity of the tool and the workpiece at temperatures that are high enough to cause softening. Here, the maximum temperature has been simulated to be near the melting point of the tool piece and the temperature distribution is also illustrated.

4.3.2   FEM Plots of AA-6061 Alloy

4.3.2.1   Von-Mises Plot

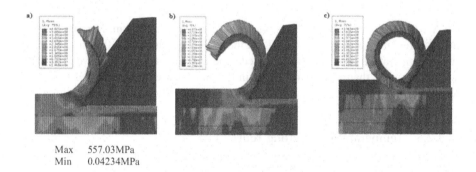

Max   557.03MPa
Min   0.04234MPa

FIGURE 4.8   Von-Mises stress at (a) 0.25 ms, (b) 0.5 ms and (c) 0.75 ms.

4.3.2.2   Temperature Plot

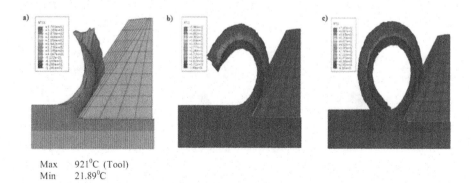

Max 921^{0}C (Tool)
Min 21.89^{0}C

FIGURE 4.9   Shear stress at (a) 0.25 ms, (b) 0.5 ms and (c) 0.75 ms.

4.3.2.3 Shear Stress Plot

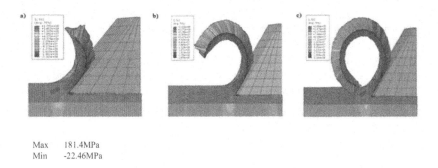

Max   181.4MPa
Min   -22.46MPa

FIGURE 4.10   Shear stress at (a) 0.25 ms, (b) 0.5 ms and (c) 0.75 ms.

4.3.2.4 Shear Strain Plot

In AA-6061, the chip formation is observed in a continuous form. Maximum Von-Mises stress is observed as 557.3 MPa at the tool–chip interface at 0.95 ms. A maximum temperature of 921°C was reached. The maximum strain observed is 0.607, which is higher than what is observed in the AA-2024 due to its higher CTE at a comparable high temperature that leads to a higher plastic strain. This is illustrated in Figures 4.8–4.11.

4.3.2.5 Temperature on Cutting Tool

Finite element analysis reveals that the maximum temperature observed on cutting tool is 921°C, whereas the melting point of steel 4340 is above 1493°C. Hence, the tool wear is expected to be lesser than machining the Al-2024 alloy. The process of shear stress distributionis observed to be maximum on the chip and the tool–chip interface after cutting. Infrared measurement of the temperature at the tool–chip interface while machining Ti-6Al-4V has been demonstrated and documented by Heigel et al. (2017). This investigation relies on the infrared thermographic imaging method to measure the temeprature rise and profiles of

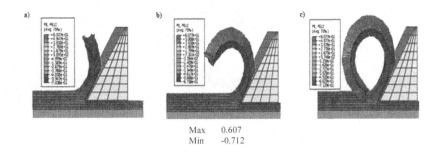

Max   0.607
Min   -0.712

FIGURE 4.11   Shear strain at (a) 0.25 ms, (b) 0.5 ms and (c) 0.75 ms.

the tool–workpiece interface and nearby areas in machining. The infrared thermography method depends only on the radiative content of the heat generation and not the small but neglected portion of heat relased, which can only be detected through convection and tool to workpiece conduction through the contact area. We recommend a hybrid technique that considers an intelligent locating of thermocouples coupled with infrared thermography camera. Hence, Dubovska and Majerik (2015) and Heigel et al. (2017) report temperture detection that are suspected to be lower than the total generated temperature. Though the temperature measurements are reported to be precise, the simulated temeratures are always seen to be higher than the measured temeratures experimentally, for reasons mentioned in the Results and Discussion section.

4.3.2.6 Von-Mises Stress and Chip Formation

Figure 4.12 represents the surface after machining. We can observe the no-red regions on the workpiece that indicate less stress. Another plot represents the chip formed after machining. If we observe the chip, it is continuous in form as compared to the previous simulation with AA-2024. This is due to the absence of strain localization which is otherwise influenced by discontinuities in plastic strain at higher temperatures that result in softening. It has also been experimentally observed. Kalyan and Samuel (2015) have conducted a carefully designed cutting-mode analysis in high-speed finish turning of an AlMgSi alloy using edge-chamfered PCD tools. They have used high-cutting speeds and low-feed rates for their study. The depth of cut is different at 0.5 mm with a high shear friction coefficient of 0.8. However, the intial surface roughness conditions of the workpiece are not reported. Some of the findings in normal turning are seen to be supportive of the simulations in orthogonal turning like the shear stress simulations where the values are understandably high. The alloy used by them is closer in composition and manufacturing method to the 6061 alloy simulated here.

FEM plots of AA-7075 alloy

4.3.2.7 Von-Mises Plot

In AA-7075, the chip formation is observed as a continuous form; although, in the beginning a small pinch of chip is seen to fly away. Maximum Von-Mises stress for

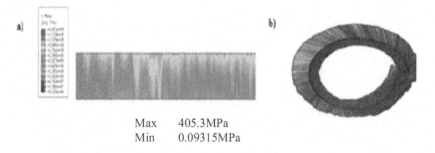

| | Max | 405.3MPa |
| | Min | 0.09315MPa |

FIGURE 4.12   (a) Von-Mises stress distribution on workpiece after machining. (b) Chip formed during machining simulation.

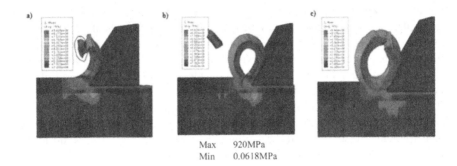

Max   920MPa
Min   0.0618MPa

FIGURE 4.13   Workpiece Von-Mises stress at (a) 0.25 ms, (b) 0.5 ms and (c) 0.75 ms.

the workpiece is observed as 916.9 MPa at the tool–chip interface at 0.95 ms. This is represented in Figures 4.13 and 4.14, where the maximum tool temperature is also noticed to be high.

4.3.2.8   Temperature Plot

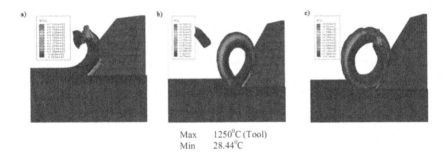

Max 1250^0C (Tool)
Min 28.44^0C

FIGURE 4.14   Temperature at (a) 0.25 ms, (b) 0.5 ms and (c) 0.75 ms.

4.3.2.9   Shear Stress Plot

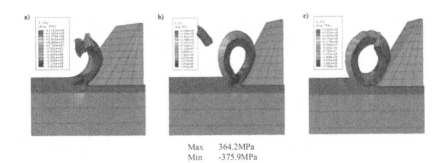

Max   364.2MPa
Min   -375.9MPa

FIGURE 4.15   Shear stress at (a) 0.25 ms, (b) 0.5 ms and (c) 0.75 ms.

4.3.2.10   Shear Strain Plot

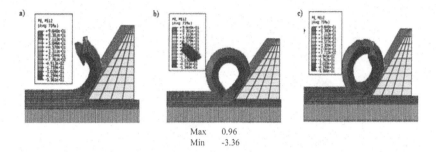

Max   0.96
Min   -3.36

FIGURE 4.16   Shear strain at (a) 0.25 ms, (b) 0.5 ms and (c) 0.75 ms.

4.3.2.11   Temperature on Cutting Tool

Figures 4.15 and 4.16 show shear stress and strains induced in the workpiece during machining. The maximum temperature observed on the cutting tool is 1250°C, whereas the melting point of 4340 steel is 1493°C. Hence, the tool wear will be appreciable, but the wear will be in between that of the values obtained while machining Al-2024 or Al-6061. The tool edge also experiences plastic deformation. The shear stress distribution of the process is observed to be maximum on the tool–chip interface. It is observed that shear stress and strains are higher than the two previous cases due to the exponents n and m being higher for the AA-7075 alloy. The material constants A, B and D_3 are also high compared to the other two alloys (see Tables 4.1–4.4). The experimental analysis of the cutting forces obtained in dry turning processes of UNS A 97075 aluminum alloys is well documented in De Augustina (2013). This alloy is the same as the 7075 alloy used in this investigation for simulation. For our orthogonal turning, the cutting-force evaluations for three alloys are given at the end of our discussion.

4.3.2.12   Von-Mises Stresses and Chip Formation

The surface of workpiece after machining has higher stress regions than AA-2024 or AA-6061 workpieces as indicated by numerical values of shear stresses though no red regions are seen. This is because of a different color coding for shear stress values in this plot. Figure 4.17 shows surfaces with severe residual stresses. The figures represent the chips formed during machining. The first one is the continous chip formed after cutting, and the second chip is the pinch formed during the first 0.25 ms of operation. It has been observed that a significant adiabatic temperature rise leads to discontinuities in thermal properties and causes strain localization, which reduces the chips to pinch after an incremental time frame. Another factor seen to affect the turning temperture rise is illustrated in an experimental study on the correlation between turning temperature rise and turning vibration in dry turning on Al alloys (Yu et al., 2019). This study experimentally proves why the detected interface and tool tip temeperatures can be affected by vibrations observed during dry turning operation. However, our investigation on the turning simulation has not considered any vibration. But it is noteworthy that while comparing our

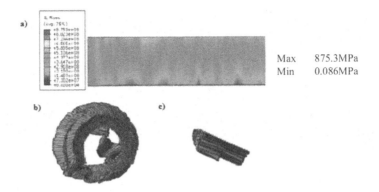

Max   875.3MPa
Min   0.086MPa

FIGURE 4.17   (a) Von-Mises stress distribution on the AA-7075 workpiece after machining, (b) continuous chip and (c) pinched chip

simulations with the temperature rise observed experimentally by others, vibrational effects are naturally accounted for in those temperature measurements that might be lower due to discontinuities. This is yet another reason why the Johnson-Cook simulation results in higher temperatures than the experimnetally observed values.

The cutting forces for all three aluminum alloys, the orthogonal turning process of which was simulated, were generated based on the assumed depth of cut, length of turning and the shear stress simulated. From shear stress plots for three alloys AL-2024, 6061 and 7075 given in Figures 4.4, 4.10, 4.15, respectively, the maximum cutting forces along the circumferential direction were generated as mentioned next.

The AL-2024 produced a cutting force of 496 N, the AL-6061 produced a cutting force of 362 N and the AL-7075 produced a cutting force of 728 N after 2mm length of turning operation as their shear stresses were different. This is seen to be comparable to the turning cutting forces reported in the ASM handbook chapter for our assumed alloys and machining parameters (ASM Handbook, Volume 16, 1989). Rakotomalala and Caperra (2004) give a general overview of the simulation of orthogonal machining and damage laws concerned with chip formation.

4.3.3 Discussion and Summary

The book by Paulo Davim is seen to be very influential as a practical guide in the understanding of the machining processes in general (Paulo Davim, 2011). In summary from our simulations on orthogonal turning and the supporting investigations, it has been observed that:

1. AA-2024 is a stronger and stiffer material with a high-induced stress and predicted tool wear after the operation.
2. AA-6061 is good for turning and the induced temperature and expected tool wear are less.
3. AA-7075 experiences high stresses during turning. The temperature experienced by the cutting tool and the tool wear are high. So tool wear is expected to be more in turning of AA-7075.

4. During the cutting of AA-2024, the initial cutting temperature is the highest; in AA-6061, it is moderate; and in AA-7075, it is high for the chosen turning parameters.

5. In AA-7075, the relatively brittle nature is somewhat more than in AA-2024, so the chips are noticed to be discontinuous at first. The AL-2024 alloy is less brittle, so it produces continuous chipping.

6. AA-6061 is ductile in nature at the turning temerature so it has continuous chipping during the turning process. So AA-6061 is mainly used in aerospace applications, but AA-2024 has less density. AA-2024 is preferred for sheet metal operations rather than machining operations despite the density advantage.

7. The maximum observed localized strains increase from AA-2024 to AA-6061 and then exhibit a higher value in the AA-7075 alloy. In all three cases, strain localization was seen to occur in the chip area as evinced from the plots and play a role in the looping behavior of the chips. Strain localization was also seen to decide whether the chip would be continuous or discontinuous based on the banding that were seen to occur as a result of the strain localization. The banding occurs differently in all three materials due to localization effects, and the associated surface phenomena, such as chip discontinuity or continuity, due to adiabatic heating. Strain localization occurs due to the rake angle and the cutting-edge angle selection, which causes the looping of the chips during formation and growth. Intitially, strain localization is formed as a result of the chip–tool interface thermomechanical phenomena due to the development of shear stresses and strains that manifest later along the length of the chip. Differentials are attributed to differing thermal properties and John-Cook constants of the chip and tool materials. The coefficient of friction in three cases is also reponsible for this effect.

8. The adiabatic heat rise and temperature recorded as a result of the simulation based on John-Cook model does not consider the discontinuities that occur in the thermal conductivity or diffusivity of the tool or the workpiece near the softening or melting temperatures. These discontinuities that occur practically reduce the maximum observed temperature which otherwise could not be considered here in the simulation based on the model. Many properties are actually discontinuous variables with respect to rise in temeprature and not constant as projected in the available limited data given in Tables 4.1–4.4 of this investigation. Hence, the tool piece or the workpiece achieve the simulated temperatures in milliseconds, which is practically not observed. Though the recorded heat rise and temperatueres are slightly exceeded in simulation, the time taken to reach the maximum temeprature is slightly longer. This is one of the limitations of the model that actually has to be constructed based on a few additional relationships that arise from changes in temperature during the defined operation. These could be material-specific aiding in decision making for choice of materials and processes.

9. Structural polymorphic, isomorphic and allotropic changes in a workpiece caused by thermal effects and rate of heating and cooling due to turning form a separate topic and require more tools to analyze. They are considered to be outside the purview of this investigation. It is also important to look at the

material as a long-term patient rather than as a workpiece. Many researches on machining as only a productivity tool ignore these facts raised in points 7, 8 and 9, in particular, researches by a particular group (Dumitrescu et al., 2006; Jahanmir, 2011; Jahanmir et al., 2018). These researchers seem to concentrate more on optimizing productivity through tools and approaches by varying the machining parameters rather than considering the short- and long-term effects of these optimization excercises on the workpiece material. The workpiece is rather a victim than a patient. Hence, improving the efficiency of existing material models like the John-Cook model to suit the needs of SHM (structural health monitoring) of machined products is the need of the hour. Here, the nature of chipping and the resultant thermomechanical effects can be predicted better with the aid of a non-equillibrium phase diagram and a time-temperature transformation diagram which is simply impossible with the Taguchi, Grey relational, orthogonal arrays and ANOVA approaches that do not consider metallurgical effects but mainly concentrate on the material removal rate, chipping, chatter, tool wear and surface cracking. A descriptive thesis has been presented by Deepa et al. on machining-induced structural changes in self-reinfocred composites and its influence on the mechanical properties. Here, both the conventional and unconventional machining techniques were studied for the molecular influence and modification in the mechanical properties (Deepa et al., 2020; Deepa et al., 2020a; Deepa 2021). This aids in decision-making processes for choice of materials and methods of manufacturing.

10. From the values obtained from the machining (orthogonal turning) simulations, comparisons were made between all the three aluminum alloys by using summary graphs for easy grading and decision making, as shown in Figures 4.18 and 4.19.

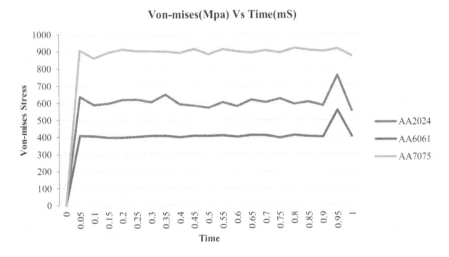

FIGURE 4.18   Von-Mises (MPa) versus time (ms) plot for three aluminum alloys based on FEM studies.

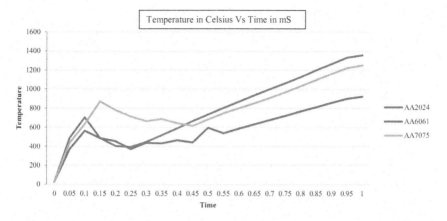

FIGURE 4.19   Temperature (°C) versus time (ms) plot for three aluminum alloys based on FEM studies.

4.4   CONCLUSIONS

The main aim of this contribution concerns the comprehension of physical phenomenon accompaying chip formation, maximum Von-Mises stresses, shear stresses, shear strain and maximum temperature developed in three different aluminum alloys during turning by using the John-Cook model and damage law. The main outcome of this work is in introducing the FEM methodology to explain thermal aspects of the turning process of three important aluminum alloys using an original approach of coupled thermal-explicit dynamic analysis based on the John-Cook model. A time step path-dependent Lagrangian analysis approach was followed. Assumptions based on this model are found to be useful in the prediction of the extent of adiabatic heating, chip removal processes, chip discontinuities and pinching, Von-Mises equivalent stresses, shear stresses and strains, margins of thermal safety, cutting forces and tool reliability. A comparison of the finite element simulation of orthogonal turning of three different popularly used aluminum alloys is seen to produce similarities to results obtained from actual turning trials reported by other workers. The simulation results on temperature profiles were understandably higher than the experimental measurements. The model is useful in assisting in the choice of materials and process selection for selected applications. Limitations of the model are also spelt out in the discussion that points out to the change in the relationship between properties and parameters as a function of temperature.

ACKNOWLEDGEMENT

Authors thank the management of VIT, Vellore, for the encouragement and support.

FUNDING

This project was funded by the School of Mechanical Engineering, VIT, Vellore, India.

CONFLICT OF INTEREST

Authors state that there is no conflict of interest with any other group.

REFERENCES

Asad, M., Ijaz, H., Khan, M.A., Mabrouki, T. and Saleem, W. (2014). Turning modeling and simulation of an aerospace grade aluminum alloy using two-dimensional and three-dimensional finite element method. *Proceedings of the Institution of Mechanical Engineers, Part B: Journal of Engineering Manufacture*, 228: 367–375.

Bharathi Raja, S. and Baskar, N. (2011). Particle Swarm optimization technique for determining optimal machining parameters of different workpiece materials in turning operation. *Journal of Advanced Manufacturing Technology*, 54: 445–463.

Bhoyar, M.Y.R. and Kamble, P.P.D. (2013). Finite element analysis on temperature distribution of turning process. *International Journal of Modern Engineering Research*, 3: 541–546.

Cakir, A., Yagmur, S. and Sekar, U. (2016). The effect of minimum quantity lubrication under different parameters in the turning of AA 7075 and AA 2024 aluminium alloys. *Journal of Advanced Manufacturing Technology*, 84: 2515–2521.

Daoud, M., Chatelain, J.F. and Bouzid, A. (2015). Effect of rake angle on Johnson-Cook material constants and their impact on cutting process parameters of aluminium 2024-T3 alloy machining simulation. *Journal of Advanced Manufacturing Technology*, 81: 1987–1997.

De Augustina, B., Bernal, C., Camacho, A.M. and Rubio, E.M. (2013). Experimental analysis of the cutting forces obtained in dry turning processes of UNS A 97075 aluminium alloys. *Procedia Engineering*, 63: 694–699.

Deepa, A. (2021) Influence of Conventional and Unconventional Machining Induced Structural Changes on the Mechanical Characterization of Self Reinforced Polymer Composites. Doctoral thesis, SMEC, VIT University, India.

Deepa, A., Kuppan, P. and Krishnan, P. (2020). A comparative study of structural changes in conventional and unconventional machining and mechanical properties evaluation of polypropylene based self reinforced composites. *Science and Engineering of Composite Materials*, 27(1): 108–118.

Deepa, A., Padmanabhan, K. and Kuppan, P. (2020a). Impact of polymeric changes on mechanical and structural properties of polyethylene self reinforced polymer composites. *U.P.B. Sci. Bull., Series D*, 82(3): 179–194.

Dubovska, R. and Majerik, J. (2015). Experimental measurement of temperature in turning AlCu3MgMnPb aluminium alloy. *Recent Advances in Mechanical Engineering and Automatic Control*, pp. 122–125.

Dumitrescu, P., Koshy, P., Stenekes, J. and Elbestawi, M.A. (2006). High-power diode laser assisted hard turning of AISI D2 tool steel. *International Journal of Machine Tools and Manufacture*, 46: 2009–2016.

Evdokimov, D.V. et al. (2014). Thermal stress research of processing and formation of residual stress when end milling of work piece. *World Applied Sciences Journal*, 31: 51–55.

Frager da Silva, T. and Reis, A. (2017). Simulation studies of turning of aluminium cast alloy using PCD tools. *Procedia CIRP*, 58: 555–560.

Heigel, J.C., Whitenton, E., Lane, B., Donmez, M.A., Madhavan, V. and Moscoso-Kingley, W. (2017). Infrared measurement of the temperature at the tool-chip interface while machining Ti-6Al-4V. *Journal of Materials Processing Technology*, 243: 123–130.

Ijaz, H., Zain-ul-abdein, M., Saleem, W., Asad, M. and Mabrouki, T. (2017). Modified Johnson-Cook plasticity model with damage evolution: Application to turning simulation of 2XXX aluminium alloy. *Journal of Mechanics*, 33: 777–788.

Jahanmir, S. (2011). Surface integrity in ultra high speed micromachining. *Procedia Engineering*, 19: 156–161.

Jahanmir, S., Tomaszewski, M.J. and Heshmat, H. (2018). Ultra high-speed micro-milling of aluminum alloy. *Advances in Multidisciplinary Engineering-ASME*. DOI: 10.1115/1.861080_ch18.

Javidika, M., Sadhegifar, M. and Jahazi. M. (2020) On the impact of tool geometry and cutting conditions in straight turning of aluminium alloys 6061-T6; an experimentally validated numerical study. *Journal of Advanced Manufacturing Technology*, 106: 4547–4565.

Kalyan, C. and Samuel, G.I. (2015) Cutting mode analysis in high speed finish turning of AlMgSi alloy using edge chamfered PCD tools. *Journal of Materials Processing Technology*, 216: 146–159.

Klocke, F. (2011). *Manufacturing Processes 1—Cutting*, Springer, Switzerland, pp. 197–212.

Klocke, D.L.F. and Puls, H. (2013). FEM-Modelling of the thermal workpiece deformation in dry turning. In *14th CIRP Conference on Modeling of Machining Operations (CIRP CMMO)*.

Kumar, R., Pattnaik, S.K. and Sarangi, S.K. (2019). Influence of cutting parameters on cutting forces and surface roughness in dry turning of aluminium using PCD and different coated steels. *Sadhana*, 44: 186.

Mabrouki, T., Girardin, F. and Rigal, J. (2008). Numerical and experimental study of dry cutting for an aeronautic aluminium alloy (A2024-T351). *International Journal of Machine Tools & Manufacture*, 48: 1187–1197.

Krenzer J. Machining of aluminium and aluminium alloys. (1989). *Metals Handbook*, Volume 16, ASM, Metals Park, Ohio, pp. 761–804.

Markopoulous, A.P. (2012). *Finite Element Method in Machining Process*, Springer, Switzerland.

Osorio-Pinzon, C.J., Abolghasem, S. and Rodriquez, J.P.C. (2019). Predicting the John-Cook constitutive model constants using temperature rise distribution in plane strain machining. *Journal of Advanced Manufacturing Technology*, 105: 279–294.

Panatale, J.O., Rakotomalala, R. and Caperra, S. (2004). 2D and 3D numerical models of metal cutting with damaged effects. *Computer Methods in Applied Mechanics and Engineering*, 193: 4383–4399.

Paszta, P. (2017). Modelling the aluminium alloy machining process. MATEC Web of Conferences, CoSME'16, Poland, 94: 02010.

Paulo Davim, J. (2011). *Modern Machining Technology—A Practical Guide*. Woodhead Publishing, UK.

Rudrapati, R., Sahoo, P. and Bandhyopadhyay, A. (2016). Optimization of process parameters in CNC turning of aluminium alloy using hybrid RSM cum TLBO approach, IOP Conf. Series. *Materials Science and Engineering A*, 149: 012039.

Santos, M.C. Jr., Machado, A.R. and Ezugwu, E.O. (2015). Multiobjective optimization of cutting conditions when turning aluminium alloys 1350-O and 7075-T6 grades using genetic algorithm. *Journal of Advanced Manufacturing Technology*, 76: 1123–1138.

Sawarkar, N. and Boob, G. (2004). Finite element based simulation of orthogonal cutting process to determine residual stress induced. *International Journal of Computer Applications*, 0975-8887: 33–38.

Shrot, A. and Baker, M. (2012). Determination of Johnson-Cook parameters from machining simulations. *Computational Materials Science*, 52: 298–304.

Stakhniv, N.F. and Devin, L.N. (2019). The influence of the cutting speed on the temperature and forces at the precision turning of non-ferrous metals using cutters with round diamond hard-alloyed plates. *Journal of Super Hard Materials*, 41: 128–184.

Xiao, X., Abhushawashi, Y. and Astakhov, V.P. (2011). FEM simulation of metal cutting using new approach to model chip formation. *International Journal of Advances in Machining and Forming Operations*, 3: 71–92.

Yu, Q., Li, S., Zhang, X. and Shao, M. (2019). Experimental study on correlation between turning temperature rise and turning vibration in dry turning on Al alloy. *Journal of Advanced Manufacturing Technology*, 103: 453–469.

Section III

Robotics

5 Development of Forward Kinematics of a Two-DOF Manipulator in MATLAB® Environment

Ranjan Kumar, Kaushik Kumar
Birla Institute of Technology, Ranchi,
Jharkhand, India

CONTENTS

5.1 INTRODUCTION

From the historical perspective, it can be summarized that continuous efforts have been dedicated towards technological growth in the last few decades to fulfill the demands and to overcome the complexities involved in various manufacturing processes. As a result, the field of robotics has emerged magnificently. For decades, humans are fascinated about developing the "human-like creature" that can perform the assigned tasks efficiently using some level of intelligence. Researchers have studied this field very efficiently in the last few years and have developed a useful and lucid explanation for the same. Also, continuously emerging new technologies are being added from time to time in this field of robotics to enhance clarity and their efficient use. Human life in their workplace is associated with much complexities, uncertainties, and unmodeled environments where it is difficult to work. In such situations, smart and intelligent machines play a significant role, and intelligent robots get deployed in such areas to perform tasks where human feels uncomfortable and difficult to work. Performing heavy-duty works, lifting high payloads, house cleaning, performing various complex tasks in factories as well as in houses, a land survey in unknown places, healthcare systems, transportations, and space explorations are some of the examples [1–5]. However, the robotic system is nowadays an essential part of our rapid developmental arena but incorporating the robotic system has increased the fear and insecurities of losing their jobs among people. The feeling of hatred towards robots still exists today and the wide range of application arena of robotic systems in various fields has been retarded effectively [3]. However, the very first in the 1940s, Isaac Asimov envisioned robots as human helpers in his science fiction stories and postulated certain laws of robotics [6].

The mechanisms have been developed and been studied since the invention of wheels, which was the very first kind of mechanism found in the archaeological excavation [7] and the development has gone through various stages. Decades of developments have proved the existence of mechanisms from manual to automated machines. Many technical pieces of literature and thoughts have been summarized to develop different types of mechanisms as per requirements. Watt and Stephenson mechanisms [8] as shown in Figure 5.1 are one such example of their kind [9]. Around the mid-twentieth century, researchers focused mainly on the investigation of the human and machine association, and this was the time when people started working with the automated mechanisms for performing their specified jobs along with paving the path for new opportunities and developments. The era of the 1960s was the period of historical developments in the field of industrial automation characterized by rapid technological growth and changes in popular methodologies. These automated machines were identified as the unique devices of that time which were termed "industrial robots" [10]. After having a long-time development, nowadays the robotics play a very vital role and has become an essential part of our life [11–12] as because we the human being are known for designing and developing products as per our need with suitable cost and quality that is essential in enhancing our production rates. Further, the new technological growth has led to design of collaborative systems [13], keeping the safety concern of the human workers with the autonomous

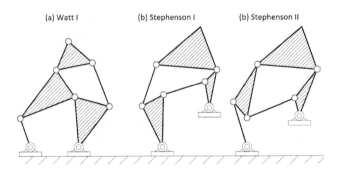

FIGURE 5.1   Schematic depicting the Watt and Stephenson mechanisms [8].

machines or robots. The sense of collaboration between the robot and the human has opened-up opportunities for many technological solutions such as (i) enhanced production rates, (ii) ergonomic concerns, (iii) meeting up the demands of variable products, (iv) beating the market fluctuations, and, most important, (v) safety measures in the workplace. These are also much helpful towards new complexities involved in the human–machine interactions [14–17]. Industrial robots are mainly employed in risky and dangerous environments, which have certainly undergone serious risk assessments. Nowadays, in order to enhance the productivity and to meet market demands, industries are undergoing robotization. Automotive industries are one of the major examples and the perfect area of robotization.

Robots are classified according to their specifications which depends mainly on the requirement and their applications that certainly varies from purpose to purpose [18–19]. There are many types of robots available as per their specified jobs and specifications which are generally classified as parallel and serial manipulators [20]. Serial manipulators are the most commonly used manipulators that are formed by having a series of connections of rigid bodies known as links [21]. These links are serially connected together by means of joints as shown in Figure 5.2.

Another important point to be noted that industrial robots used in industries may or may not contain some obstacles in their configuration space or working area which is required to undergo the path planning or trajectory planning process so that the collision with any kind of obstacles can be avoided during their operation time. The planning of such a collision-free path is done by a predetermined set of points [22] in the cartesian space for the static environment or in the dynamic environment where the path continuously changes.

"In both cases, these sets of points should be transformed from a configuration space to joints space utilizing of inverse kinematics" [23]. Hence, industrial robots are being introduced rapidly across the world among various industries and along with having reliable data referring to their rapid growth during the last few years. One can refer to the statistics of the world robotics report released by the International Federation of Robotics (IFR). The subject of robotics is a multidisciplinary domain of study and requires a great amount of computational work. This computational work is sometimes very tedious and time-consuming. So, to overcome the computational

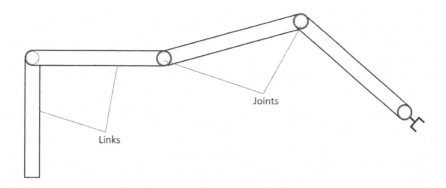

FIGURE 5.2   The schematic of a serially connected manipulator.

complexities and for a time-saving approach, we need to have any computational platform or any third-party software which can provide us the computational help in a much lesser time with accurate and reliable results. Here, MATLAB® does a tremendous job as its name suggests. MATLAB® provides a large number of software packages for multiple applications with high-performance numerical computation and visualizations and is known for its one of the most widely used engineering tools nowadays. Its interactive environments featured with many built-in functions made this a very much sophisticated software package to perform technical computations, graphical representation, and animation. MATLAB® provides easy extensibility with its own high-level programming language [24].

5.2   KINEMATIC TRANSFORMATIONS

As we know that a robotic manipulator has to perform tedious, unpredicted, and unmodeled work in a complex environment with many unknown facts and to perform these jobs, it is essential for the manipulator to follow a predetermined path or trajectory or it is necessary to define a set of instructions for a particular trajectory for a particular robotic system or manipulator. In order to follow such a specific path in the configurational space, it is essential to have "control over various joints and positions of each link" of the manipulator to follow an accurate trajectory for the desired result. A robotic system possesses multiple rigid bodies called links and each individual link is connected serially by means of joints. Also, "the degree of freedom (DOF) of a robot depends upon the number of associated links and joints, their types, and the kinematic chain." Generally, in 3-D cartesian space, a rigid body has six degrees of freedom (DOF), and if there occurs "no relative motion between two or more links then from the kinematic views it is said to have a single link." The proper knowledge of the "kinematics, mechanisms and dynamic modelling" of a system is much helpful in designing and developing a robotic manipulator. The development of the kinematics model of a manipulator is important in providing clear visualization and information relating to the pose of the end effector in 3-D cartesian space. The kinematics information is necessary for getting "kinematics of motion" and the "higher-order derivatives of the position describe the differential motions" in terms of velocity and acceleration which is independent of forces or torques applied at the joints of the manipulator [25–30].

Reileaux [31] is renowned as the "Father of Kinematics" because of his significant contributions in the domain of kinematics and machine theory. The main insight of his kinematic theory was that he observed a machine's mechanisms as a chain of restricted parts and estimated "the existence of geometrical constraints between the kinematic pairs." Reileaux [32] described some essential symbolic notations to represent the number of variables required to offer a thorough kinematic description of the system. "Kinematic description, kinematic property, and the system analysis" were the major focus of his idea of kinematics and this became the basis of the "synthesis of new mechanisms of a system." However, the proposed notations were insufficient in describing the kinematic systems, and Denavit-Hartenberg [33] noted this shortcoming and proposed a rethinking of the "kinematic symbolic notation, which completely defines kinematic properties using the lower-pair via kinematical equations."

Furthermore, the main focus of a mechanism is to allow the relative motion and transition of this motion from one part to another across the machine parts.

The "slider-crank mechanism," where the linear motion of the piston-cylinder arrangement is turned into a continuous circular motion, is one of the best examples of such motion transformation. By using the existing surface contacts between the pieces, the connections among neighboring parts are exploited for producing relative motions. As a result, the primary goal of establishing a link between relative pairs is achieved in a certain proportion so that the desired output motion can be produced in response to an input motion. The two mating parts that make up the pair are referred to as a "pair," and each individual part of such mating pairs is referred to as an "element of the pair." The occurrence of surface contacts between mating pair elements is referred to as a "lower Pair," and when these contacts are confined to a line, it is referred to as a "higher pair." The lower pair of a mechanism is explored first because of their relative simplicity. "Spherical pair, cylindrical pair, simple pair, screw pair, revolute pair, and sliding pair are examples of possible lower pairs."

Whittaker [34] have reported on several of Euler and Chasles' "kinematical theories and geometrical proofs, as well as many other theorems, and has traced their work." The chapter's main goal is to present a large picture of a rigid sequential body's positions in 3-D space using "linear transformations or, in some cases, both translational and rotational transformations." The work of Paul [35], Suh and Radcliffe [36], and Beggs [37] can be accessed and noted because the study of finite displacement of a rigid body is essential for obtaining the pose of a rigid body and by computing solutions to the transformation of a rigid body in 3-D space and finding the relative positions and orientations by using rotational and translational parameters. Laub [38] proposed that the pose of a rigid body in 3-D space can be calculated by designating three fixed non-collinear points in the rigid body. "The motion of a rigid body can be estimated and the sequence of displacements can be represented in the matrix form using a series of composite transformations. The provided approach can describe the pose of a rigid body even in a simpler manner from the imprecise data obtained from the measurement of the positions of the three non-collinear points."

5.3   SPATIAL DESCRIPTION

The spatial description possesses the pose of a rigid body. The pose comprises the "position and orientation" of the rigid body in 3-D configurational space. The motion of the rigid body is described in terms of translation and rotation in 3-D cartesian space. the translation of a rigid body includes three cartesian coordinates, whereas the rotation includes their angular coordinates. Thus, all these six coordinates are responsible for the complete description of a rigid body. In the study of the kinematics of a robot manipulator, we have to study the "pose" or the "configuration" of various rigid bodies or links with respect to some frame of reference. The "position" and "orientation" of the manipulator is described with respect to the **base coordinate system**, also termed as **fixed coordinate system** or **universal coordinate system**, that is, {U}. Also, the position and orientation of the object is described with the help of the center of mass and one other coordinate system attached to it, called the **body coordinate system**, that is, {B} as shown in Figure 5.3.

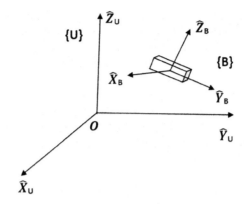

FIGURE 5.3 The schematic diagram showing a body in space with the attached coordinate frame.

For having the complete description of a moving frame, that is, the body coordinate frame, {B}with respect to the fixed reference frame, that is, the universal coordinate frame {U}, the location and the direction of the axes of {B} must be specified. This is specified by describing a vector between the origin of the moving frame {B} and the origin of the reference frame {U}. Therefore, the moving frame can be expressed by three vectors describing the orientation of the body frame or moving frame and the remaining fourth vector describing the position of the body frame or moving frame and is given as:

$$[T] = \begin{bmatrix} r_{11} & r_{12} & r_{13} & P_x \\ r_{21} & r_{22} & r_{23} & P_y \\ r_{31} & r_{32} & r_{33} & P_z \\ 0 & 0 & 0 & 1 \end{bmatrix} \tag{5.1}$$

However, a moving frame {B} can also be represented by a 3×4 matrix without having the scale factor which is not generally common. Therefore, a fourth row having the scale factor is added to the matrix to make the given matrix a 4×4 homogenous matrix.

5.3.1 HOMOGENOUS TRANSFORMATION MATRIX

The homogenous transformation matrix is necessary during matrix operations for various reasons, such as, the very first reason is that during a matrix operation, it is easier to calculate the inverse of the square matrix than rectangular matrices. Second, having the same dimensions of the matrices is necessary for the multiplication of two matrices. It means the number of rows (i.e., m rows) in the first matrix should be equal to the number of columns (i.e., n columns) in the second matrix for getting the matrix multiplication of $m \times n$ matrix. Also, the first matrix dimension of $m \times n$ multiplied with matrix dimension $n \times p$, results the matrix of dimension $m \times p$.

If the two square matrices P and Q having same dimension $(n \times n)$ and $(n \times n)$, then their multiplication gives the same results of dimension $n \times n$. So, if we multiply P by Q or Q by P, both give a square matrix having same dimension. But, if the two matrices R and S having different dimensions are multiplied gives the RS matrix multiplication but of different dimension other than the dimensions of R and S. Here, it is important to note that R can be multiplied by S but S may or may not be multiplied by R and the result RS have the different dimension other than R and S.

5.4   REPRESENTATION OF FRAME TRANSFORMATIONS

The transformation is nothing but simply moving in space freely with respect to a fixed frame of reference. When a moving frame or an object having its own body coordinate system {B} moves in space with respect to another fixed frame of reference or universal coordinate system {U, then the movement of the object or the moving frame is represented or described by the concepts of transformations which comprises both translation and rotation of the object in space. The moving frame has its own location and orientation change in space, so it can be said that the transformation is the change of state of the moving frame. The transformation can be defined in either of the following three ways:

    i. pure translation,
    ii. pure rotation, and
    iii. combination of both translation and rotation.

5.4.1   TRANSLATION OF A FRAME

Let's suppose there is a moving frame, that is, {B} is translated with respect to the fixed frame or the universal frame of reference, that is, {U} as shown in Figure 5.4. The position of the origin of the moving frame {B} is represented by a position vector $^U\mathbf{r}_{orgB}$ with respect to the universal frame of reference {U}. Since there is only the translation involved here, so axes of both frames will be parallel, that is,

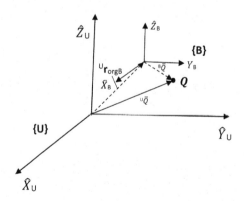

FIGURE 5.4   The schematic showing the frame translation in space [24].

axes $\hat{X}_U$, $\hat{Y}_U$, and $\hat{Z}_U$ of the universal frame is parallel to the axes of the moving frame of reference $\hat{X}_B$, $\hat{Y}_B$, and $\hat{Z}_B$. Now, let's suppose the position of any point Q with respect to the body coordinate system is known to be $^B\overline{Q}$. Then, the location of the point Q can be estimated with respect to the universal frame of reference and can be given as:

$$^U\overline{Q} = {}^U\mathbf{r}_{\text{orgB}} + {}^B\overline{Q} \tag{5.2}$$

Now let's suppose the origins of both the moving coordinate system and the fixed coordinate system are coinciding with each other and then the frame is rotated anti-clockwise with respect to the universal frame of reference and the point Q is as it is in the space which position vector $^B\overline{Q}$ is known with respect to moving frame {B}, then in such cases, the position of point Q in the fixed frame of reference {U}, that is, $^U\overline{Q}$ and is represented as:

$$^U\overline{Q} = {}^U R_B + {}^B\overline{Q} \tag{5.3}$$

Here, no translation occurs; hence, the translation of origin represented by the position vector $\mathbf{r}_{\text{orgB}}$ vanishes.

Further, suppose if there occurs both translation and rotation of the moving frame {B}, that is, the moving frame {B} is translated by $^U\mathbf{r}_{\text{orgB}}$ from {U} and has also been rotated anti-clockwise by an angle about the universal frame {U} as shown in Figure 5.5. Also, if the position of the point Q with respect to the body coordinate system {B} is known, then in such case, the $^U\overline{Q}$ is represented as:

$$^U\overline{Q} = {}^U R_B \,{}^B\overline{Q} + {}^U\mathbf{r}_{\text{orgB}} \tag{5.4}$$

Here, this is important to note that the translation and rotation, that is, the position and orientation can be represented together is termed as "transformations" and

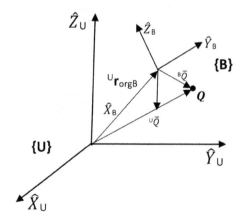

FIGURE 5.5 The schematic showing a combined form of transformation (translation and rotation) [24].

is represented in terms of transformation matrix $[T]$. Hence, the previous equation can be represented as:

$$^U\overline{Q} = {}^U T_B \, {}^B\overline{Q} \tag{5.5}$$

As we already know that the transformation matrix $[T]$ is a 4×4 homogenous transformation matrix in which the first three columns show the rotation or orientation and the last column represents the position and the fourth row of this homogenous transformation matrix contains the elements 0,0,0 and 1 which are used just to modify and compensate the matrix dimensions as given in Equation (5.1). The homogenous transformation matrix is represented symbolically as:

$$[T] = \begin{bmatrix} R_B^U(3X3) & \vdots & r_{orgB}^U(3X1) \\ \cdots & \vdots & \cdots \\ 0 \quad 0 \quad 0 & \vdots & 1 \end{bmatrix} \tag{5.6}$$

Here, it is important to note that the first three elements of the fourth row represent the **perspective transformation** and the fourth element of this fourth row represents the **scaling factor**. Also, note that in the **cartesian coordinate system** (CCS), the rotation matrix or the orientation can be expressed in terms of **roll, pitch,** and **yow; Euler angles** and others. Similarly, the position of a body in space can be represented in **cartesian, spherical,** and **cylindrical coordinate system.** Hence, from Equation (5.1), the pure translation (without any rotation) can be given by a 4×4 transformation matrix as shown:

$$[T] = \begin{bmatrix} 1 & 0 & 0 & P_x \\ 0 & 1 & 0 & P_y \\ 0 & 0 & 1 & P_z \\ 0 & 0 & 0 & 1 \end{bmatrix} \tag{5.7}$$

Here, it can be seen that in case of pure translation, the matrix contains 3×3 rotation matrix which is an identity or a unit matrix. The translation of a body or a plane is denoted by **TRANS** operator and is represented as **TRANS** $(\hat{X}, P)$ (indicates the direction and magnitude of translation) and is given as:

$$TRANS(\hat{X}, P) = \begin{bmatrix} 1 & 0 & 0 & P_x \\ 0 & 1 & 0 & 0 \\ 0 & 0 & 1 & 0 \\ 0 & 0 & 0 & 1 \end{bmatrix} \tag{5.8}$$

The translation operator is commutative in nature, that is,

$$\textbf{TRANS}(\hat{X}, P_x)\textbf{TRANS}(\hat{Y}, P_y) = \textbf{TRANS}(\hat{Y}, P_y)\textbf{TRANS}(\hat{X}, P_x) \tag{5.9}$$

Also, the translation along all the three axes can be written in shorter form by using a single **TRANS** operator as:

$$\mathbf{TRANS}(\hat{X}, P_x)\mathbf{TRANS}(\hat{Y}, P_y)\mathbf{TRANS}(\hat{Z}, P_z) = \mathbf{TRANS}(P_x, P_y, P_z) \qquad (5.10)$$

Similarly, the pure rotation (without any translation) can be given by the transformation matrix using Equation (5.1) as:

$$[T] = \begin{bmatrix} r_{11} & r_{12} & r_{13} & 0 \\ r_{21} & r_{22} & r_{23} & 0 \\ r_{31} & r_{32} & r_{33} & 0 \\ 0 & 0 & 0 & 1 \end{bmatrix} \qquad (5.11)$$

The rotation of a body or a frame is represented by **ROT**, which indicates the direction of rotation about an axis and angle of rotation and is represented as **ROT** $(\hat{Z}, \theta)$. Hence, the rotation about Z-axis by an angle θ (theta) is represented by a 3×3 rotation matrix and is given as:

$$ROT\,(\hat{Z}, \theta) = \begin{bmatrix} \cos \theta & \sin \theta & 0 \\ -\sin \theta & \cos \theta & 0 \\ 0 & 0 & 1 \end{bmatrix} \qquad (5.12)$$

And, corresponding to Equation (5.12), the 4×4 transformation matrix can be written as:

$$[T] = \begin{bmatrix} \cos \theta & \sin \theta & 0 & 0 \\ -\sin \theta & \cos \theta & 0 & 0 \\ 0 & 0 & 0 & 0 \\ 0 & 0 & 0 & 1 \end{bmatrix} \qquad (5.13)$$

Similarly, the rotation about Y-axis, that is, ROT $(\hat{Y}, \theta)$ and the rotation about Z-axis, that is, ROT $(\hat{Z}, \theta)$ can also be given as:

$$ROT\,(\hat{Y}, \theta) = \begin{bmatrix} \cos \theta & 0 & \sin \theta \\ 0 & 1 & 0 \\ -\sin \theta & 0 & \cos \theta \end{bmatrix} \text{and,} \qquad (5.14)$$

$$ROT\,(\hat{X}, \theta) = \begin{bmatrix} 1 & 0 & 0 \\ 0 & \cos \theta & -\sin \theta \\ 0 & \sin \theta & \cos \theta \end{bmatrix} \qquad (5.15)$$

The rotation matrix is not commutative in nature, that is,

$$\mathbf{ROT}(\hat{X}, \theta_1)\mathbf{ROT}(\hat{Y}, \theta_2) \neq \mathbf{ROT}(\hat{Y}, \theta_2)\mathbf{ROT}(\hat{X}, \theta_1) \tag{5.16}$$

It is to be noted that each row or each column of a rotation matrix is considered as a unit vector because the magnitude of each row or column of the rotation matrix is equal to 1 separately. Also, one can note that the dot product of each row with other row equals to 0 and this is also true for each column with other column of the matrix. One more illustration is very much important to be noted that in terms of rotation matrix, the inverse of rotation matrix gives the transpose of it.

$$\mathbf{ROT}^{-1}(\hat{X}, \theta) = \mathbf{ROT}^{T}(X, \theta) \tag{5.17}$$

Further, sometimes it is required to obtain the final transformation of a frame with respect to a fixed reference frame, it has to go under certain number of successive translation and rotation sequence. "Any transformation can be resolved into a set of translations and rotations in a particular order." For example, we can give a rotation to a frame about x-axis and can translate it along x and y axis and again we can rotate this along z-axis to obtain the final transformation as per our requirement. It is important to note that the order is very important to be followed here. If we change the order of two successive transformations then we will get entirely different result. For better picture, let's summarize the successive transformation of a rotating frame with respect to a fixed frame of reference as:

    i. rotation about x-axis by an angle θ_1,
   ii. followed by a translation relative to x and y by a distance l_1 and l_2,
  iii. followed by rotation about z-axis by an angle θ_2.

Let's suppose a point q having coordinates (n, o, a), that is, q_{noa} is attached to the origin of the moving frame. Now, as the frame rotates or translates, the attached point q_{noa} also rotates or translates with respect to some fixed frame of reference and the coordinates of the point also changes relative to the reference frame. So, after taking the first transformation of rotation about x-axis by an angle θ_1, the coordinates of point q relative to the fixed frame of reference can be calculated as:

$$q_{1,xyz} = \mathbf{ROT}(\hat{X}, \theta_1) \times q_{noa} \tag{5.18}$$

Similarly, after the second transformation, the coordinates of the point relative to the fixed frame of reference can be calculated as:

$$q_{2,xyz} = \mathbf{TRANS}(l_1, l_2) \times q_{1,xyz} = \mathbf{TRANS}(l_1, l_2) \times \mathbf{ROT}(\hat{X}, \theta) \times q_{noa} \tag{5.19}$$

Again, after having final transformation of rotation about z-axis, the coordinates of the point relative to the reference frame can be given as:

$$q_{xyz} = q_{3,xyz} = \mathbf{ROT}(\hat{X}, \theta_2) \times q_{2,xyz} = \mathbf{ROT}(\hat{X}, \theta_2) \times \mathbf{TRANS}(l_1, l_2) \times \mathbf{ROT}(\hat{X}, \theta) \times q_{noa}$$

(5.20)

It can be noticed here that the final transformation relative to the frame of reference can be obtained by pre-multiplying the coordinates of the point by each transformation matrix. Also, it is important to note that the final transformation is obtained by reversing the order of transformation taking place [24].

5.5   FORWARD AND INVERSE KINEMATICS

As we have already discussed that a robot has to perform a specified assigned task with high degree of accuracy and efficiency. For obtaining the desired results, the information related to the position and orientation, that is, the pose of the end-effector of a manipulator is required. The pose of the manipulator relative to a reference frame is essential in obtaining the solutions related to the position problem. By differentiating once and twice the position problem, the velocity and acceleration problem is obtained for analysis of a robot. This information is important in terms of motion control of the end-effector and the dynamic analysis of the robotic system. The information related to the pose of a manipulator is obtained by the analysis of robot kinematics, which includes the forward kinematics and the inverse kinematics. In "forward kinematics" of a manipulator, the joint angles or the joint variables are known and we try to find the position and orientation of the end effector based on this information. Note that the position and orientation we obtained here is not the desired pose, but it has been only obtained based on the known information of joint variables, which is not really required in the real-life scenario. Now suppose, we want our robot arm to be at our desired location or position and with desired orientation, then in such cases, we must have to find the required joint angles for which our robot could reach at the desired position with desired orientation. Hence, finding the joint variables based on the desired position of the end effector is known as "inverse kinematics" of a robot. In reality, it is necessary to find the inverse kinematics using the equations of forward kinematic so that the robot controller can estimate the required joint angles and can run the robot accordingly to reach at the desired position. So, the first step is to determine the equations of forward kinematics and then using these equations, we can find the inverse kinematics of a robot manipulator. The position and orientation, that is, the pose of a rigid body in space is completely defined by six DOF. In other words, six pieces of information are required for estimating the position and orientation of a rigid body in a 3-D cartesian space. Therefore, in case a robotic arm, we must need to attach a coordinate frame to the robot hand and the position and orientation information of the robot hand is determined. A schematic multi-link robotic manipulator with attached coordinate frame and their relative position and orientation has been delineated in Figure 5.6. However, there occurs many possibilities for the same configurations. Also, the representation of the final desired pose of a robotic system, basically, depends upon

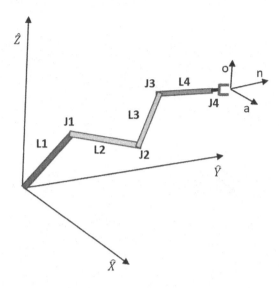

FIGURE 5.6   The schematic of a multi-link robotic arm with an attached coordinate at the end effector.

types of coordinate frames we are choosing for the robot motion. A robot can be chosen to move in space whose position can be defined in the "cartesian coordinate frame" and its orientation can be represented using the roll pitch yaw (RPY) set of joints. Similarly, a robot can also be chosen to be designed in a spherical coordinate frame for its positioning and Euler angles for its orientation. Hence, the calculation of forward and inverse kinematics for such situations are quite complicated. Hence, instead of going for such complicated design, the Denavit-Hartenberg representation of robot kinematics is quite efficient and recommended.

5.6   DENAVIT-HARTENBERG KINEMATIC NOTATIONS

A paper was published in *ASME Journal of Applied Mechanics* authored by Denavit-Hartenberg [26] in the year 1955. Later on, this becomes the basis for robot modelling and its representation. The paper describes a useful technique to derive the equation of motion for various robot manipulators. Nowadays the Denavit-Hartenberg notation or often termed as D-H parameters has become the best-suited technique to define the robotic motion regardless of the level of complexities involved. The D-H notations are also useful in representing the various transformation taking place in any of the coordinate systems which we have already discussed like "cartesian, cylindrical, spherical, Roll Pitch Yow (RPY) and Euler." This is also useful in representing the possible occurring combinations of links and joints involved in a robot manipulator, such as SCARA robot and revolute articulated robot. Preparation of the D-H is necessary as it added up some benefits in terms motion calculation, Jacobian, dynamic analysis, force or torque estimation as well as in trajectory planning [9, 27–31].

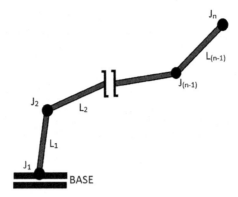

FIGURE 5.7   The schematic diagram of n-link serial manipulator with n-joints and $(n-1)$ links.

Robots are made up of the combination of multiple links and joints such as "revolute (rotation) joint or prismatic (linear) joint," which can move either in the same plane or in different planes. Also, there may be some offset having the link length as zero or of some definite length, sometimes having twisted or bent in nature lying on any plane. Hence, any general set of links and joint combinations can be considered as a robot model regardless of whether it is following any such coordinate systems or not, which we have discussed previously. Therefore, our only aim is to model and analyze such robotic systems with basic principles of DH notations. A serial manipulator with rigid link joints can be seen in Figure 5.7.

Hence, to model and analyze a robotic system we need to assign the coordinate frame at each joint of the model. This is a general procedure for obtaining the transformation from base to first joint, from second to third joint and so on until we reach at the final joint. The combination of each of these joint transformations gives the final or overall transformation of a robot model. The overall transformation matrix $^{n}T_{0}$ for the robotic system as depicted in Figure 5.7 indicates the pose of a point located at the last joint (J_{n}) with regard to the base joint (J_{1}). By convention, it is important to note that the coordinate frame attached to "the end-effector should be at the central position of the tool. This position is called the Tool Centre Point (TCP)." In addition to that, each joint of a manipulator connects the two neighbor links. Hence, if a robotic system is having n joints, then there will be $(n-1)$ links in the manipulator. The general procedure of obtaining the overall transformation of a model has been further summarized in next DH representation section.

5.6.1   DH Parameters

For kinematics analysis of a robot manipulator, some important rules and parameters have been defined by Denavit-Hartenberg. These are called link and joint parameters which follow some standard set of rules as mentioned next. The schematic view of link and joint parameters can be observed in Figure 5.8.

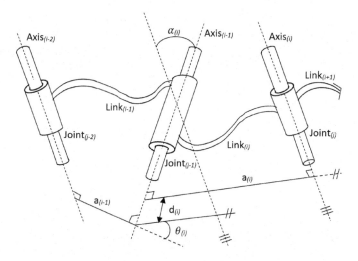

FIGURE 5.8   The schematic diagram showing links and joint parameters [24].

Link parameters, namely the link length (a_i) and the angle of twist (α_i) are basically used to define the complete structure of a manipulator as discussed next:

- **Length of Link (a_i):** Link length is a mutual perpendicular distance between the axis$_{(i-1)}$ of the joint$_{(j-1)}$ and the axis$_{(i)}$ of the joint$_{(j)}$.
- **Angle of Twist (α_i):** It is defined as the angle between axis$_{(i-1)}$ and axis$_{(i)}$.

Similarly, joint parameters such as link offset (d_i) and joint angle (θ_i) are useful in describing the positions of the neighboring links of a manipulator.

- **Offset of Link (d_i):** It is the measurement of the distance from the point of intersection between $a_{(i-1)}$ and axis$_{(i-1)}$ and to the point of intersection between a_i and axis$_{(i-1)}$ and is measured along the said axis.
- **Joint Angle (θ_i):** It is the measurement of joint variable or angle which is an extension of $a_{(i-1)}$ and a_i, measured about the axis$_{(i-1)}$.

It is important to note that in case of revolute joint, the joint angle (θ_i) is considered as a variable and all other parameters remain constant. Whereas, in case of prismatic joint, the link offset (d_i) is variable and all other parameters remain constant. The previously defined four parameters are making our task quite easy to assign the coordinate frame at the joint of a manipulator and using the mechanisms involved in DH notation, the determination of forward kinematics is absolutely straightforward.

In the schematic diagram shown in Figure 5.8, it can be analyzed for the associated Denavit-Hartenberg links and joint parameters. For a simple understanding, we can summarize that from "O," we have to reach at "B." So, while moving from

"O" to "B" via "A," we have to go through various transformation such as translations and rotations. In this way, the final and overall transformation, that is, $^{i-1}T_i$ can be given as follows:

$$^{i-1}T_i = {}^{i-1}T_O {}^O T_A {}^A T_B {}^B T_i \tag{5.21}$$

$$^{i-1}T_i = \mathbf{ROT}(\hat{Z}, \theta_i)\mathbf{TRANS}(\hat{Z}, d_i)\mathbf{ROT}(\hat{X}, \alpha_i)\mathbf{TRANS}(\hat{X}, a_i) \tag{5.22}$$

$$^{i-1}T_i = Screw_{(Z)}Screw_{(X)} \tag{5.23}$$

$$^{i-1}T_i = \begin{bmatrix} \cos\theta_i & -\sin\theta_i\cos\alpha_i & \sin\theta_i\sin\alpha_i & a_i\cos\theta_i \\ \sin\theta_i & \cos\theta_i\cos\alpha_i & -\cos\theta_i\sin\alpha_i & a_i\sin\theta_i \\ 0 & \sin\alpha_i & \cos\alpha_i & d_i \\ 0 & 0 & 0 & 1 \end{bmatrix} \tag{5.24}$$

Example

A schematic of a two-link manipulator having two links and two joints has been shown in Figure 5.9. The two joints are nothing but the revolute joint, so the joint angle will be the variable here and all other remaining parameters are constant. Also, the link length of the manipulator is L_1 and L_2 [24].

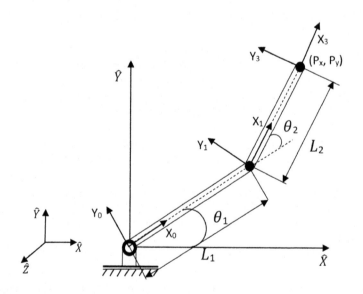

FIGURE 5.9   Two link manipulator with attached coordinate frame [24].

TABLE 5.1
DH Table for a Two-link Manipulator

Frame	θ_i	d_i	α_i	a_i
1	θ_1	0	0	L_1
2	θ_2	0	0	L_2

Using the link and joint parameters, the DH table can be drawn as provided in Table 5.1.

The overall homogeneous transformation matrix can be written for this two-link manipulator as follows:

$$^{Base}T_2 = {}^{Base}T_1 {}^1T_2 \tag{5.25}$$

By the use of DH table, the first transformation of link 1 relative to base (i.e., $^{Base}T_1$) can be computed as:

$$^{Base}T_1 = \mathbf{ROT}(\hat{Z}, \theta_1)\mathbf{TRANS}(\hat{X}, L_1)$$

$$^{Base}T_1 = \begin{bmatrix} c_1 & -s_1 & 0 & L_1c_1 \\ s_1 & c_1 & 0 & L_1s_1 \\ 0 & 0 & 1 & 0 \\ 0 & 0 & 0 & 1 \end{bmatrix} \tag{5.26}$$

Here, c_1 and s_1 denotes cos (θ_1) and sin (θ_1), respectively.

Similarly, the second transformation of link 2 relative to link 1 (i.e., $^1T_2$) can be obtained as:

$$^1T_2 = \mathbf{ROT}(\hat{Z}, \theta_2)\mathbf{TRANS}(\hat{X}, L_2)$$

$$^1T_2 = \begin{bmatrix} c_2 & -s_2 & 0 & L_2c_2 \\ s_2 & c_2 & 0 & L_2s_2 \\ 0 & 0 & 1 & 0 \\ 0 & 0 & 0 & 1 \end{bmatrix} \tag{5.27}$$

Therefore, the overall homogeneous transformation matrix can be computed using Equation (5.25) and can be given as:

$$^{Base}T_2 = {}^{Base}T_1 {}^1T_2$$

$$^{Base}T_1 = \begin{bmatrix} c_{12} & -s_{12} & 0 & L_1c_1 + L_2c_{12} \\ s_{12} & c_{12} & 0 & L_1s_1 + L_2s_{12} \\ 0 & 0 & 1 & 0 \\ 0 & 0 & 0 & 1 \end{bmatrix} \tag{5.28}$$

Here, from this homogeneous transformation matrix given in Equation (5.28), one can observe that the "position of the end effector," after performing the overall transformation according to the defined DH table, can be given as:

$$P_{(x,y,z)} = \begin{bmatrix} P_x \\ P_y \\ P_z \end{bmatrix} = \begin{bmatrix} L_1c_1 + L_2c_{12} \\ L_1s_1 + L_2s_{12} \\ 0 \end{bmatrix}$$

Also, the orientation of the end effector with regards to the base coordinate frame can be observed by the rotation matrix shown in the previous given homogeneous transformation matrix as:

$$R_{(x,y,z)} = \begin{bmatrix} c_{12} & -s_{12} & 0 \\ s_{12} & c_{12} & 0 \\ 0 & 0 & 1 \end{bmatrix}$$

The following is the sample MATLAB® code that one can write to estimate the "forward kinematics of any DOF system robot manipulator." However, here we have demonstrated MATLAB® code [39] for a two-link or two-DOF manipulator as shown in Figure 5.9. As we are demonstrating only the forward kinematics of a manipulator, it means we may have the known values of joint variables (i.e., θ_1 and θ_2) for the manipulator, and we will try to find the position of the manipulator in the configuration space. So, in the sample code we will ask the user to enter the values of joint variables for both revolute joints.

The given sample code (Figure 5.10) has been performed in MATLAB® for a case of simple two-link manipulate. Now the similar procedure can be adopted in coding the multi-link manipulator. However, the function file is coded in a separate new script file and this function file must be saved using the function name which we have chosen for our function. Here, in this case, we have created two function files, one for homogeneous transformation matrix as "HT_matrix" and, second, for the line joining between the coordinate points as "join_line." The user can input the value of joint variables theta_1 and theta_2 in the command window and the user can see the designed manipulator in a separate window.

The homogeneous transformation matrix function, that is, "HT_matrix" function file and join_line function file can be coded as.

Similarly, the "join line" function file can be coded in a separate script file to join the coordinate points obtained in space by a line. The function file must be saved with the function name that has been created as shown in Figure 5.11 of the sample MATLAB® code.

Further, from the command window, one can see the results of individual transformations by fetching the HT_1 and HT_2. Also, the overall transformation that are being taking place can be accessed by putting the Total_HT in the command window. However, MATLAB® is a mathematical software and it follows some basic rules of programming that can be accessed as a tutorial from MATLAB® online documentation for beginners. One can program in MATLAB® in their own fashion or he/she can edit this program as per the best of their use.

```
clear all;
clc;
%format G
```

Defining Links and Joints Parameters) %%

DH parameters for First Link

```
joint_1 = input('Enter theta1: ');   % First joint angle (Angle of movement)
d1 = 0;                               % link offset_1    (Distance b/w two x)
alpha_1 = 0;                          % twist angle_1    (Angle between two z)
a1 = 200;                             % link length_1    (perpendicular dist b/w two z)

% DH parameters for First Link
joint_2 = input('Enter theta2: ');   % Second joint angle
d2 = 0;                               % link offset_2
alpha_2 = 0;                          % twist angle_2
a2 = 200;                             % link length_2
```

Defining Homogeneous Transformstion

```
HT_1 = HT_matrix(joint_1,d1,alpha_1,a1); % for homogeneous transformation i.e. HTM 1
HT_2 = HT_matrix(joint_2,d2,alpha_2,a2); % for homogeneous transformation i.e. HTM 2
TH = HT_1*HT_2;                          % overall  homogeneous transformation i.e. HTM

zeroth_cord = zeros(4);
first_cord = HT_1;
second_cord = HT_1*HT_2;

join_line(zeroth_cord,first_cord);

hold on
grid on
join_line(first_cord,second_cord);   % joining the coordinates by line
axis([-350 350 -300 300 0 150])      % 3D Axis interval
xlabel('X - Axis');
ylabel('Y - Axis');
zlabel('Z - Axis');
legend('First Link','Second Link')
title('Two Link Robot Manipulator')
```

FIGURE 5.10 Sample MATLAB® code for input and outputs.

```
function [DH_param] = HT_matrix(theta, d, alpha, a)
% This HTM function is to solve the forward kinematics problem:
DH_param = ...
[cosd(theta), -1*sind(theta)*cosd(alpha), sind(theta)*sind(alpha), a*cosd(theta);...
 sind(theta),  cosd(theta)*cosd(alpha), -1*cosd(theta)*sind(alpha), a*sind(theta);...
 0, sind(alpha), cosd(alpha), d;...
 0, 0, 0, 1];
end
```

FIGURE 5.11 Sample MATLAB® code for function file.

5.7   RESULT AND DISCUSSION

This work has been summarized by showing the application of MATLAB® programming in the field of robotics in a simple and lucid way, which can be understood by anyone interested in the subject. However, some prior knowledge of the basics of kinematics and transformations of rigid bodies in space is necessary which also have been summarized in this chapter. The result is showing that MATLAB® is nothing but a sophisticated mathematical tool that can be used to meet the extensive calculation requirements in our day-to-day research work, academics purposes, and so on. MATLAB® helps us to program our mathematical model in an efficient and easy manner to obtain the desired results. MATLAB® is available with many powerful toolboxes almost in every field of work and research with lots of in-built function which helps the user in many ways for easy results. However, the main approach in this work is to familiarize the user with the basics of MATLAB® implementation according to the theory- or subject-based knowledge with the help of basic MATLAB® programming language. More or less, the concept behind any programming language is similar and one can work in MATLAB® having the basic knowledge of any programming language, like C, C++, Python, and Java. These programming languages provide the efficient use of MATLAB® software by using in-built functions and ready-made command available in each toolbox. Hence, the conceptual understanding of MATLAB® toolboxes and hands-on practice is necessary for their efficient use. In case of robotics, it is essential to have an in-depth insight of the subject matter and proper understanding of equations involved in kinematics and dynamics of these machines. Only then one can achieve the mathematical model using MATLAB® and other platforms in a fast, efficient, and easy way of robot modeling.

5.8   CONCLUSION

In the present chapter, we have tried to implement MATLAB® environment tool through programming approach in the development of a forward kinematics of a basic two-link robot manipulator for educational purpose. The main purpose of this work is to give basic insights to users of MATLAB® in the very multidisciplinary field of robotics. The chapter has used the basic concepts of robot kinematics and transformations of a rigid body in space on the basis of a popular Denavit-Hartenberg notations as discussed earlier. However, the way of implementation of the basics of kinematics and transformations is not the only way but a user can understand these concepts and can apply in their own way based on the basic programming language and built-in MATLAB® functions and commands. Apart from the programming approach discussed in this chapter, one can use the in-built robotics toolbox provided in MATLAB®. The robotics toolbox has been provided with many interesting features that supports wide variety of robot prototypes available in its library. The toolbox has many interesting features, such as DH representation of a manipulator, capabilities of visualizing the position, velocity, and acceleration of various joints, path planning, trajectory planning of a manipulator, dynamics, and control design of various manipulators can be done in an efficient manner. MATLAB® also provides a simulation approach of a robot manipulator using Simulink. MATLAB® Simulink is one of the powerful built-in systems that one can use to design a mathematical model using equations and can analyze responses. Development of direct kinematics, inverse kinematics, control dynamics of manipulator, graphical user interface to design the controller of a manipulator can be awarded as the best features of MATLAB® software. Further, one of the major toolboxes designed by Peter Corke can also be used in robotic studies. The toolbox has in-built library of various industrial robotic systems which can also be visualized in 3-D, the realistic behavior and responses of which can be seen by entering the appropriate command.

5.9 FUTURE SCOPE

The present study is limited to developing the forward kinematics of a two-link robot manipulator through MATLAB® programming environment. This can be further processed for developing the forward kinematics of multi-link manipulators and the same can also be done for the development of inverse kinematics of a manipulator, since robotics is a multidisciplinary research area carrying various possibilities in different working arena. The programming through MATLAB® or any other programming language can be introduced for obtaining such results. Furthermore, the same work can be explored for developing the path planning of such manipulators via forward and inverse kinematics approach. Since a manipulator has to perform various jobs at unwanted and unpredictable environments, therefore, in this regard, the accuracy of path following that depends upon the joints angle obtained through inverse kinematics becomes necessary and to meet the desired accuracy, the introduction of optimization approach becomes essential. Various heuristics and metaheuristics optimization approach can further be introduced for solving the kinematics problems and path planning of robot manipulators instead of traditional approaches. Such introduction of optimization algorithms leads the domain of robotics development to another level of designing the intelligent control systems. Such intelligent control techniques follow some characteristics like "self-learning, quick adaptation, self-organizing, self-diagnosis etc." The use of such intelligent algorithms is a part of artificial intelligence and machine learnings that has become one of the essential parts of today's development in robotics field. Such development is very useful in designing and developing the intelligent expert control system of today's era. These have become possible due to advanced intelligent computing techniques which are required to be investigated in depth. This in-depth investigation is the necessity of today's modern control design in order to avoid associated flaws and limitations to improve the sophistication and robustness of the system.

REFERENCES

[1] R. C. Dorf. 1988. *International Encyclopaedia of Robotics*. New York: John Wiley & Sons.
[2] Robot Shop Distribution Inc., History of Robotics: Timeline. 2008. [Online]. Available http://www.robotshop.com/media/files/PDF/timeline.pdf.
[3] R. K. Mittal and I. G. Nagrath. 2019. *Robotics and Control*. Chennai: McGraw Hill Education.

[4] Ranjan Kumar and Kaushik Kumar. 2021. Intelligent control design schemes of a two-link robotic manipulator. *Artificial Intelligence in Mechanical and Industrial Engineering*, Chapter 5. Boca Raton: CRC Press. https://doi.org/10.1201/9781003011248.

[5] R. Shukla Tiwari and R. Kala. 2010. *Real Life Applications of Soft Computing*, 1st edition. CRC Press. https://doi.org/10.1201/EBK1439822876.

[6] S. K. Saha. 2013. *Introduction to Robotics*. New Delhi: McGraw Hill Education.

[7] J. N. Postgate. 2017. *Early Mesopotamia: Society and Economy at the Dawn of History*. London: Taylor & Francis. https://doi.org/10.4324/9780203825662.

[8] M. M. Plecnik and J. M. McCarthy. 2016. Design of Stephenson linkages that guide a point along a specified trajectory. *Mechanism and Machine Theory*, 96(1), 38–51. https://doi.org/10.1016/j.mechmachtheory.2015.08.015.

[9] G. S. Soh and N. Robson. 2013. Kinematic synthesis of minimally actuated multi-loop planar linkages with second order motion constraints for object grasping. *Proceedings of the ASME 2013 Dynamic Systems and Control Conference (DSCC 2013)*, pp. 1–8. https://doi.org/10.1115/DSCC2013-4029.

[10] J. J. Craig. 2005. *Introduction to Robotics: Mechanics and Control*, 2nd edition. Singapore: Addison-Wesley.

[11] M. Daneshmand, O. Bilici, A. Bolotnikova, and G. Anbarjafari. 2017. Medical robots with potential applications in participatory and opportunistic remote sensing: A review. *Robotics and Autonomous Systems*, 95, 160–180. https://doi.org/10.1016/j.robot.2017.06.009.

[12] N. Spolaôr and F. B. V. Benitti. 2017. Robotics applications grounded in learning theories on tertiary education: A systematic review. *Computers and Education*, 112, 97–107. https://doi.org/10.1016/j.compedu.2017.05.001.

[13] R. Kumar and K. Kumar. 2021. Intelligence-assisted cobots in smart manufacturing, Chapter 2. *Advanced Computational Methods in Mechanical and Materials Engineering*, 1st edition, pp. 19–41. Boca Raton: CRC Press.

[14] V. Gopinath and K. Johansen. 2016. Risk assessment process for collaborative assembly – A job safety analysis approach. *Procedia CIRP*, 44, 199–203. https://doi.org/10.1016/j.procir.2016.02.334.

[15] G. Michalos, S. Makris, J. Spiliotopoulos, I. Misios, P. Tsarouchi, and G. Chryssolouris. 2014. ROBO-PARTNER: Seamless human-robot cooperation for intelligent, flexible and safe operations in the assembly factories of the future. *Procedia CIR*, 23, 71–76. https://doi.org/10.1016/j.procir.2014.10.079.

[16] T. Brogardh. 2007. Present and future robot control development – An industrial perspective. *Annual Reviews in Control*, 31(1), 69–79. https://doi.org/10.1016/j.arcontrol.2007.01.002.

[17] J. Kruger, R. Bernhardt, and D. Surdilovic. 2006. Intelligent assist systems for flexible assembly. *CIRP Annals – Manufacturing Technology*, 55(1), 29–32. https://doi.org/10.1016/S0007-8506(07)60359-X.

[18] A. Kramberger, A. Gams, B. Nemec, D. Chrysostomou, O. Madsen, and A. Ude. 2017. Generalization of orientation trajectories and force-torque profiles for robotic assembly. *Robotics and Autonomous Systems*, 98, 333–346. https://doi.org/10.1016/j.robot.2017.09.019.

[19] S. Makris, P. Tsarouchi, A.-S. Matthaiakis, A. Athanasatos, X. Chatzigeorgiou, M. Stefos, K. Giavridis, and S. Aivaliotis. 2017. Dual arm robot in cooperation with humans for flexible assembly. *CIRP Annals*, 66(1), 13–16. https://doi.org/10.1016/j.cirp.2017.04.097.

[20] M. W. Spong, S. Hutchinson, and M. Vidyasagar. 2005. *Robot Modelling and Control*. New Delhi, INDIA: John Wiley.

[21] H. N. Ghafil, A. H. Mohammed, and N. H. Hadi. 2015. A virtual reality environment for 5-DOF robot manipulator based on XNA framework. *International Journal of Computer Applications (IJCA Journal)*, 113(3), 33–37. https://doi.org/10.5120/19808-1601.

[22] R. Mathew and S. S. Hiremath. 2016. Trajectory tracking and control of differential drive robot for predefined regular geometrical path. *Procedia Technology*, 25, 1273–1280. https://doi.org/10.1016/j.protcy.2016.08.221.

[23] V. N. Iliukhin, K. B. Mitkovskii, D. A. Bizyanova, and A. A. Akopyan. 2017. The modelling of inverse kinematics for 5 DOF manipulator. *Procedia Engineering*, 176, 498–505. https://doi.org/10.1016/j.proeng.2017.02.349.

[24] D. K. Pratihar. 2019. *Fundamental of Robotics*. New Delhi: Narosa Publishing House.

[25] C. S. G. Lee. 1982. Robot arm kinematics, dynamics and control. *IEEE Computer*, 15(12), 62–80. https://doi.org/10.1109/MC.1982.1653917.

[26] J. Luh. 1983. An anatomy industrial robots and their controls. *IEEE Transactions on Automatic Control*, 28 (2). https://doi.org/10.1109/TAC.1983.1103216.

[27] R. P. Paul. 1981. *Robot Manipulators: Mathematics, Programming, and Control*. London, England: MIT Press.

[28] M. Shahinpoor. 1987. *A Robot Engineering Textbook*. New York: Harper & Row Publishers.

[29] Y. Koren. 1985. Robotics for Engineers. New York : McGraw-Hill.

[30] K. S. Fu, R.C. Gonzalez, and C. S. G. Lee. 1987. *Robotics: Control, Sensing, Vision, and Intelligence*. New York: McGraw-Hill.

[31] Francis C. Moon and Franz Reuleaux. 2003. Contributions to 19th century kinematics and theory of machines. *Applied Mechanics Reviews*, 56(2), 261–285. https://doi.org/10.1115/1.1523427.

[32] F. Reuleaux. 1876. *Kinematics of Machinery*. Translated and edited by A. B. W. Kennedy. London: Macmillan and Company.

[33] J. Denavit and R. S. Hartenberg. 1955. A kinematic notation for lower-pair mechanisms based on matrices. *ASME Journal of Applied Mechanics*, 22(2), 215–221. https://doi.org/10.1115/1.4011045.

[34] E. T. Whittaker. 1927. *A Treatise on Analytical Dynamics of Particles and Rigid Bodies*, Chapter 1. Cambridge, MA: Cambridge University Press.

[35] B. Paul. 1963. On the composition of finite rotations. *American Mathematical Monthly*, 70, 859–862. https://doi.org/10.2307/2312674.

[36] C. Suh and C. Radcliffe. 1978. *Kinematics and Mechanisms Design*, Chapter 3. New York: Wiley.

[37] J. S. Beggs. 1966. *Advanced Mechanism*, Chapter 2. New York: Macmillan.

[38] A. J. Laub and G. R. Shiflett. March 1982. A linear algebra approach to the analysis of rigid body displacement from initial and final position data. *Journal of Applied Mechanics*, 49, 213–216. https://doi.org/10.1016/S1474-6670(17)61663-X.

[39] Rudra Pratap. 2016. *Getting Started with MATLAB: A Quick Introduction for Scientists and Engineers*, 7th edition. New York: Oxford University Press. https://global.oup.com/ushe/product/getting-started-with-matlab-9780190602062?cc=id&lang=en&.

6 Machine Learning Algorithms in Intellectual Robotics

Nithya Jayakumar
K. S. Rangasamy College of Technology, Tiruchengode,
Tamil Nadu, India

Ramya Jayakumar
Hindusthan College of Engineering and Technology,
Coimbatore, Tamil Nadu, India

CONTENTS

6.1 INTRODUCTION

Robots that can help humans in everyday settings have been a goal of Industry 4.0. The creation of robots that can perform different tasks in response to environmental context or higher-level instructions is the first step toward this aim. It is evident that a model based solely on mimicking human intelligence would be unable to replicate

DOI: 10.1201/9781003121640-9

all of the sensual and motor skills that a robot should be able to perform. In order to solve this, a complete adaptive control algorithms in the field of machine learning are proposed to train robots through trial and error.

Robots can learn from their mistakes and adapt to the situation through machine learning. Robots learn with the aid of machine learning that is similar to the gaining of knowledge by the people based on their experiences. The combination of robotics and machine learning reduces the time-consuming human training process. In industrial sector, robots can help organizations to do more tasks with fewer errors and the artificial intelligence (AI)-enabled robots perform effectively by grasping their surroundings and reacting appropriately.

The performance of robotics is enhanced to a greater extent by incorporating the concept of machine learning. A proper knowledge sharing should be done between machine learning researchers and roboticists to help the design of simulated and then finally real robotic systems. The computational powers of machine learning algorithms to boost productivity in industries have created a huge impact in recent years. Many firms have discovered that they no longer need to rely on rule-based programming, because machine learning relies on algorithms that can be learned from data. The ability of such systems to learn on their own has grown increasingly crucial in recent years, especially as organizations around the world struggle to keep up with otherwise unfathomable volumes of complicated data. As a result, machine learning is now being used in a variety of new fields. The design of fully autonomous robot intends to highlight the challenges and potential uses of incorporating learning capabilities into a robot control system, followed by a description of novel integrated system architecture. In reality, combining machine learning with optimization has been shown to significantly improve the quality of decision-making and learning ability in decision systems.

6.2   MACHINE LEARNING IN ROBOTICS

The domain of robotics is greatly influenced by the innovative and upcoming machine learning techniques. The computer vision algorithms trained by machine learning help us to build a skilled robot application. Machine learning algorithms incorporate intelligence into robots by automatically learning from data without doing any explicit programming. Machine learning requires a huge amount of data cache and needs to be taught to robots for complete learning. The learning procedure contains supervised or unsupervised algorithms and physical machines to assist robots with the machine learning process.

Intelligent robots should be able to learn through the sensor data and to behave in a continuously changing environment [1, 2]. The learning task becomes more difficult as the environment becomes more complicated. Ethologists have developed an architecture based on learning classifier systems and structural features of animal behavioral organization to address this problem. Experiments that demonstrate how behavior acquisition can be performed are offered with a discussion of the learning technique utilized and the organizational structure proposed. The virtual robot is taught to follow a light and stay away from hot and harmful objects. Harmonization is achieved through a learning coordination process, although these two simple behavioral patterns are learned independently. Sensor data should allow intelligent robots to learn how to act in a changing environment. The implementation of a humanoid walking robot is one of the important development streams in machine learning. When such talented robots are implemented in reality, the method of learning techniques will have a significant role.

6.2.1   Machine Learning in Robotics—Modern Applications

The scope of application of AI and machine learning in robotics is to help robots as follows.

  i. Recognize objects they have never seen before and recognize them in more detail
 ii. Determine a suitable position and orientation to grab an object in the environment
iii. Stay productive with active interactions and avoid any obstacle during movement
iv. Understand and prevent physical and logistic data patterns and act accordingly

6.2.1.1   Computer Vision

To build a computer vision model, developers should install a good quality camera to capture and process images by robots. The highly developed machine vision model in the field of intellectual robotics will guide the movement of robots and hence develop automated inspection systems.

6.2.1.2   Imitation Learning

The machine learning model is mainly aimed to imitate human intelligence, which is closely related to observational learning, through which newborn infants and toddlers learn. Imitation learning is used in designing robots for domains such as civil construction, farming, rescue operations during disaster and military where robotic solutions are helpful without manual intervention.

6.2.1.3   Self-Supervised Learning

Robots generate their own training samples to improve the performance using a self-supervised learning approach. This learning starts with training the sensor data collected near the robots and then to test the sensor data observed at a long range. This type of learning finds application to detect and reject objects by robots, such as identify vegetables and obstacles in rough terrain, three-dimensional scene analysis, and modeling of environment.

6.2.1.4   Assistive and Medical Technologies

The medical diagnostic data is collected and processed while assistive robots are helpful to take actions based on the medical information diagnosed. Medical robots offer diagnostic or therapeutic benefits. Surgical robots have played a tremendous role in treating patients during the COVID-19 pandemic.

6.2.1.5 Multi-Agent Learning

Intellectual robots trained using machine learning should be able to adapt to changing scenario in which the coordination and negotiation are the key aspects of multi-agent learning. This learning finds application in market-based distributed control systems where various factors are utilized in the training. Pick-and-place machines use machine learning algorithms for recognizing the object and grasp the object with the rapid processing time. Machine learning algorithms are also involved in the design of robots for unmanned operation of robots.

6.2.1.6 PCB Manufacturing

The deployment of robots during printed circuit board (PCB) design or manufacturing will complete the manufacturing process at a faster rate. Some companies are including AI technologies to boost the manufacturing process, which reduces the learning time for robots. Training people can tell what to include and what to ignore in the bot by simply clicking on the image on the screen. Trained machines also help make manufacturing more efficient by preventing confusion due to failed equipment.

6.3 MACHINE LEARNING ALGORITHMS

Robots can learn from their mistakes and adapt with credits to machine learning. People gain knowledge as a result of their experiences. In recent years, various machine learning techniques, such as supervised learning, unsupervised learning, and reinforcement learning have been used in the automation of robots [3, 4]. The machine learning model is trained through reinforcement techniques so that the robot is trained by making judgments instead of learning in a simulated environment. Different machine learning algorithms have been developed to implement autonomous operation and decision-making in intellectual robots. The expertise of the domain professionals needs to be considered in order to design and decide the machine learning model.

6.3.1 FEEDING ROBOTS WITH PLANNING AND LEARNING

AI robots gather critical strategies particularly planning and learning through machine learning. Planning is sort of teaching robots about its pace in the movement of joint to finish a venture. For example, grabbing an item with the aid of using a robotic arm is planning. Meanwhile, learning entails variety of inputs and reacts in line with facts introduced to it on an unpredictable environment. The learning process happens through various stages, such as physical demonstrations, wherein robots are trained on physical actions to be performed by them and then they are trained under various possible artificial environments. In case of supervised learning in robotics, train the model with a set of labeled values.

6.3.2 EDUCATING AND TRAINING WITH ACCURATE DATA

If the environment is static, robots can operate with predictable outcomes and preset positions. But if the environment is dynamic and unpredictable, the feedback from previous positions and environment is required for decision-making. The success of robots lies with the accuracy and volume of data used for training the model. If the training dataset is inaccurate or insufficient, robots will make wrong decisions. If the robot is trained only with minimum examples and if it is exposed to the new scenario in which it has not been trained, the robot will provide the wrong result. Thus, the robot has to be trained with sufficient training data to acquire a full potential of performance. Robots are trained to provide intellectual assistance in decision-making other than physical assistance [16, 17]. The machine learning algorithm designed for a specific robot guides the robot automatically when humans cannot monitor it. The machine learning algorithms is involved in various robotic applications like robot vision, navigation, legged locomotion, modeling vehicle dynamics, medical surgeries, and rough terrain robot navigation.

6.3.3 LEARNING

Learning algorithms [5, 6] enhances the system's performance by analyzing the changes in behavior of the learner based on the inputs and the feedback received. Complex problems that require learning algorithms are one of the most valuable tools for solving complex problems that require the intellectuality in decision-making to design machines resembling human behavior.

In the classical approach of designing robots, the robot is specifically designed to perform the required task by using the mathematical model of the robot and the environment. The parameters for the algorithm are selected manually after empirical test experiments. Whereas in the learning approach, partial parameters are initially provided and parameters are fine-tuned by the help of machine learning algorithms to acquire the desired skill. Learning controllers can automatically improve the data to attain the goal. The machine learning algorithm [9, 10] does not require an explicit program to achieve the desired goal since it allows the robot to automatically learn by experience. Based on the types of feedback that learners can interpret, learning algorithms are classified into three major categories: supervised learning, unsupervised learning, and reinforcement learning.

6.3.3.1 Supervised Learning

Supervised learning described in Figure 6.1 learns through examples of which we know the desired output. The algorithm trains the model on a labeled dataset. The sample

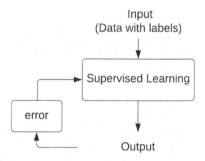

FIGURE 6.1 Supervised learning.

data and its outputs are provided as an input to the learning algorithm for its training. The process of learning and training gets repeated until a final trained model is obtained with minimum error and high accuracy. After the completion of training, the model will able to predict the accurate output for new inputs.

Algorithms of regression or classification can be used to train a robot in supervised learning. Regression algorithm whose output is a continuous variable helps to analyze the relation between the data and interprets hidden patterns, which are common among the data. Classification algorithm helps to predict the output by categories and the output of classification algorithm is discrete variable. Decision tree algorithm helps in the classification, and it is useful in 'yes or no' and 'if and then' situations. The algorithm usually mimics human thinking ability while making a decision. If it is required to solve both classification and regression problems, K nearest neighbors (KNN) algorithm will be the best choice. The algorithm works on the similarity between the data that is available and new. Based on the similarity, the new data fits into the class of similar category. The algorithm makes very accurate predictions and can compete with the most accurate model. Therefore, it can be used for applications that require high precision but not a human-readable model. The KNN algorithm does not work with high-dimensional data, and it is sensitive to noise.

The random forest algorithm is able to classify large datasets by using the number of decision trees with different decision-making options, and finally the average of all the decisions from the decision trees are taken for the final decision of the solution. This helps in providing accurate result for the complex problems as it analyzes all possible scenarios in the respective environment. It requires a lot of computational power and resources to create a large number of trees to combine the outputs.

The algorithm of support vector machine works on the creation of best decision boundary (hyperplane) which separates n-dimensional space into classes. This separation helps in classifying the new data into the appropriate category easily. The algorithm works well in high-dimensional space, and it is memory efficient, and thus can be used for face recognition, image interpretation, and text categorization in robots. The performance of the algorithm will not be good when the number of features is greater than the number of training data.

Artificial neural networks (ANNs) are the models that mimic the human brain, and it works on simple processing units called neurons. It includes an input layer, a hidden layer, and an output layer. The input layer accepts input data from the training set or test set. Each input is associated with a unique weight that is assigned based on its relative importance to other inputs. The hidden layer is the middle layer between the input layer and the output layer. The activation function is applied to the sum of the weighted inputs in this layer to recognize the pattern. The output layer returns the result of the ANN model. The use of ANN in robotics helps the robot to mimic the human behavior.

6.3.3.2 Unsupervised Learning

The real-world problems will not have expected problems and features to predict the future. Thus, unsupervised learning algorithm trains a machine for unclassified or unlabeled information and allows the machine to process the information without any supervision and prior training. The unsupervised learning algorithm

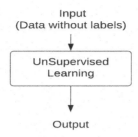

FIGURE 6.2   Unsupervised learning.

utilizes similarities, patterns, and differences between the data to group the unclassi-fied information. Thus, unsupervised learning algorithms as shown in Figure 6.2 are machine learning techniques that do not require the model to be monitored. Rather, they need to allow the model to function on its own in order to discover information.

Unsupervised learning is classified into classification and association algorithms. Clustering algorithms are the method of classification which tries to group simi-lar objects into a single cluster, whereas association algorithms work on relation between the datasets.

The unsupervised learning algorithm of K means clustering divides the unlabeled datasets into predefined clusters in which the data in each datasets have similar prop-erties. The clusters are associated with the centroid, and the distance from the cen-troid to the data decides the cluster of data. If the dimensions of the data increases, the algorithm will become complicated.

Hierarchical clustering algorithm would treat each observation as a separate clus-ter. The algorithm does not require the pre-specification of clusters. It would find two most similar clusters and merge them, thus showing the similarity between clusters. The process of merging clusters is agglomerative clustering. This step goes on itera-tively until all the clusters merge together with the main result as dendrogram.

6.3.3.3   Reinforcement Learning

Reinforcement learning [7, 8], as shown in Figure 6.3, performs learning by provid-ing rewards as a feedback to the algorithm to improve the output. Reinforcement learning can solve complex problems because of its similarity to human learning,

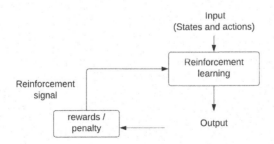

FIGURE 6.3   Reinforcement learning.

but it may be time-consuming. The reinforcement may be positive or negative. Positive reinforcement learning implements the reward for every good result in order to increase the probability of a good result, whereas negative reinforcement learning provides the penalty to improve the performance.

6.4   TRAINING MACHINE LEARNING MODELS

Machine learning helps robots to tackle unexpected environments [11–13]. The training phase in machine learning algorithm is very important as it decides the performance of the model. Among the learning algorithms, reinforcement learning is the most commonly used algorithm in robotics. In reinforcement learning, the robot initiates an initial program for doing the task. Subsequently, an algorithm is developed based upon positive and negative rewards from the unexpected environment. Thus, the robot might have learned the different task that had not been demonstrated previously [14, 15].

More number of trial process is conducted for developing the perfect algorithm in a robot. Robots are trained through demonstrations of the expert or trial-and-error approach. Significant advances in the computing power help to develop sophisticated learning algorithms in the field of robotics.

6.5   CONCLUSION

In this chapter, the application of machine learning algorithms in robotics is discussed. A well-trained, designed, and purpose-built robot can be developed in combination with machine learning algorithms, which provides absolute support to humanity. Thus, the implementation of AI and machine learning in robotics provides the solution for the use of robot in new environment where human cannot train it. In addition, robots no longer have to be built with a focus on ease of modeling but can be selected to meet task requirements for environmental sustainability, energy efficiency, and other factors. The advancement in machine learning algorithms is leading to the design of robots to solve complex industrial problems with minimum human risks.

REFERENCES

1. Alsamhi, S. H., Ma, O., & Ansari, M. (2020). Convergence of machine learning and robotics communication in collaborative assembly: Mobility, connectivity and future perspectives. *Journal of Intelligent & Robotic Systems*, 98(3), 541–566.
2. Argall, B. D., Chernova, S., Veloso, M., & Browning, B. (2009). A survey of robot learning from demonstration. *Robotics and Autonomous Systems*, 57(5), 469–483.
3. Bonaccorso, G. (2017). *Machine Learning Algorithms*. Birmingham: Packt Publishing Ltd.
4. Burrell, J. (2016). How the machine 'thinks': Understanding opacity in machine learning algorithms. *Big Data & Society*, 3(1), https://doi.org/10.1177/2053951715622512.
5. El-Shamouty, M., Kleeberger, K., Lämmle, A., & Huber, M. (2019). Simulation-driven machine learning for robotics and automation. *tm-Technisches Messen*, 86(11), 673–684.

6. Esteva, A., Robicquet, A., Ramsundar, B., Kuleshov, V., DePristo, M., Chou, K., & Dean, J. (2019). A guide to deep learning in healthcare. *Nature Medicine*, 25(1), 24–29.

7. Fong, J., Ocampo, R., Gross, D. P., & Tavakoli, M. (2020). Intelligent robotics incorporating machine learning algorithms for improving functional capacity evaluation and occupational rehabilitation. *Journal of Occupational Rehabilitation*, 30(3), 362–370.

8. Kober, J., Bagnell, J. A., & Peters, J. (2013). Reinforcement learning in robotics: A survey. *International Journal of Robotics Research*, 32(11), 1238–1274.

9. Kreuziger, J. (1992). Application of machine learning to robotics—An analysis. In Proceedings of the Second International Conference on Automation, Robotics, and Computer Vision (ICARCV'92).

10. Lesort, T., Lomonaco, V., Stoian, A., Maltoni, D., Filliat, D., & Díaz-Rodríguez, N. (2020). Continual learning for robotics: Definition, framework, learning strategies, opportunities and challenges. *Information Fusion*, 58, 52–68.

11. Murphy, R. R. (2019). *Introduction to AI Robotics*. MIT Press.

12. Pedram, S. A., Ferguson, P. W., Shin, C., Mehta, A., Dutson, E. P., Alambeigi, F., & Rosen, J. (2020). Toward synergic learning for autonomous manipulation of deformable tissues via surgical robots: An approximate q-learning approach. In 2020 8th IEEE RAS/EMBS International Conference for Biomedical Robotics and Biomechatronics (BioRob) (pp. 878–884). IEEE.

13. Peters, J. R. (2007). Machine learning of motor skills for robotics. Los Angeles: University of Southern California.

14. Peters, J., Mülling, K., Kober, J., Nguyen-Tuong, D., & Krömer, O. (2011). Towards motor skill learning for robotics. In Robotics Research (pp. 469–482). Berlin: Springer.

15. Shin, C., Ferguson, P. W., Pedram, S. A., Ma, J., Dutson, E. P., & Rosen, J. (2019, May). Autonomous tissue manipulation via surgical robot using learning-based model predictive control. In 2019 International Conference on Robotics and Automation (ICRA) (pp. 3875–3881). IEEE.

16. Siau, K., & Wang, W. (2018). Building trust in artificial intelligence, machine learning, and robotics. *Cutter Business Technology Journal*, 31(2), 47–53.

17. Wang, W., & Siau, K. (2019). Artificial intelligence, machine learning, automation, robotics, future of work and future of humanity: A review and research agenda. *Journal of Database Management (JDM)*, 30(1), 61–79.

7 Application of 3D Printing and Robotics in the Healthcare Industry

Prachi Khamkar
Next Big Innovation Labs, Bangalore, Karnataka, India

Debarshi Kar Mahapatra
Dadasaheb Balpande College of Pharmacy, Nagpur, Maharashtra, India

Atul Mansing Kadam
Ashokrao Mane Institute of Pharmacy, Ambap, Maharashtra, India

CONTENTS

DOI: 10.1201/9781003121640-10

7.1   INTRODUCTION

A technological revolution is underway in personalized patient care. Surgical templates can now be created with greater precision, and healthcare professionals can use these models to fabricate medical models for preoperative planning and patient education, as well as implants and other medical/dental devices that can improve quality of life or enhance quality of life for patients. For many patients, 3D printing (3DP) of biological tissues will lead to the fabrication of human organs in the future. 3DP, which has recently attracted attention in a range of industries, is offering new production paradigms for the fabrication of current or original product designs. Individual units of customized medications may be dispensed in a flexible manner, which is seen as a disruptive technique in the pharmaceutical industry, since it moves therapies away from one-size-fits-all dosing [1]. Because of the additive nature and the ability to integrate 3D imaging with 3DP, it is possible to use these technologies to produce 3D items that are difficult to make using standard manufacturing procedures. Since they may be used to build and evaluate creative new drug-eluting dosage forms, their usage in digital healthcare is especially exciting.

In the next three to five years, the use of robots in pharmaceutical manufacturing is likely to reach its peak. Because of its highly regulated nature, the pharmaceutical sector has been sluggish to incorporate robots and automation. Drug discovery, production, and anti-counterfeiting all benefit from collaborations between pharmaceutical businesses and suppliers of automation and robotic solutions [2]. Automation may speed up production, making it safer and more efficient, and reduce labor costs at the same time. Dispensing, sorting, kit assembly, light machine-tending, and packing are the primary functions of most pharmaceutical sector robots.

Robotics are projected to play an increasing role in personalized medicine in order to make it a reality. To deal with medicine shortages brought on by delays in the production of items like pre-filled needles, robot technology is also being used in this regard.

7.2   3D PRINTING IN PHARMACEUTICAL INDUSTRY

Since medication research and healthcare sectors are trying to minimize side effects, enhance adherence, and hence improve patient outcomes, the concept of individualized medicine has grown in popularity in recent years. 3DP has attracted a lot of attention as a possible facilitator of customized oral solid dosage forms.

As compared to withdrawing materials from a block to generate the final shape, 3DP starts from scratch and adds material only where the printer knows exactly where you want your product to be, building it up layer by layer [3]. This is essential because it enables us to produce forms and geometries that are not achievable with subtractive production.

3DP in pharmaceutical production may take one of two routes: One is the mass production of tablets, as with Spritam, and the other is the customization of dose forms for individual patients. In the past five to seven years, 3DP has grown tremendously as a study topic since it enables custom-made products. As for medicines, you may adapt the dosage and form to the patient's choice, or you can build a different geometry to guarantee that your medication can release the substance at the proper site.

7.3   TYPES OF 3D PRINTING TECHNOLOGIES IN PHARMACEUTICALS

7.3.1   SELECTIVE LASER SINTERING (SLS)

In this technique, the powder material is spread on the working platform [4]. SLS machine warms up the powder just beneath the melting temperature [5]. A laser used in this technique is carbon dioxide lasers. This helps the laser beam to draw an object by merging the powder mixture collectively and marks the object to solidify. Later on, the new layer of the powder is placed and the process rehearses again to complete the object. Approximately, 50–200 microns powder layer can be created with this technique. Finished products are left for cooling in a printing system. Materials, such as ceramics and plastic metals can be used in the SLS technique [6]. Post-treatment SLS involves limited time and energy and performance for several batches. SLS does not depend on support systems, since the unsintered powder is used in the printed surroundings of the parts. SLS printing may create complicated structures that are traditionally difficult, including fasteners or moving parts, components or loops, or other very complicated systems, such as tissue engineering and suitable for mechanical and isotropic component processing [7].

7.3.2   FUSED DEPOSITION MODELLING (FDM)

FDM is an advanced technique and categorized under the extrusion process. It is the biggest developed 3DP base in the world and is typically the first system to expose users. FDM is the widest used computer-aided design. The filament is injected into the heating block, softened, then melted with the feeder. The liquid filament is extruded from the nozzle to just above the base, and the solidification of the built-in unit's frame determines how far it travels. Successive layers are printed on the Z-plane at a distance equivalent to the thickness of the sheet through a moving platform or nozzle, which is determined by the diameter, the speed of the extrusion, and the speed of transmission of the box. Instead, it acts as a lever to push through a liquid fiber. The filament is pushed by dusting machinery. In the X-Y axis, the nozzle moves to create the first sheet on the deck base or the raft [8]. The printer begins by producing the item's outside, known as the casing, and then fills it with the percentage of infill specified. It holds to the previous layer for the cooling filament. After each sheet is finished, the surface is lowered to allow more space to be printed on a new substrate. The method will be replicated until the target is finished [9]. FDM printing permits the fabrication of complicated geometry, such as hollow pills for gastric floating. FDM 3D printers are low-cost presses that work in the same way as other 3DP systems in that filament is pushed into the heater block, softened, and then melted with the feeder. The liquid filament is extruded from the nozzle to just above the base, and the solidification of the built-in unit's frame determines how far it travels. Through a moving platform or nozzle, which is governed by the diameter, the speed of extrusion, and the speed of transmission of the box, successive layers are printed on the Z-plane at a distance proportional to the sheet thickness. Instead, it acts as a lever to push through a liquid fiber. The filament is pushed by dusting machinery. In the X-Y route, the bucket moves to create the first sheet on the deck base or raft. The printer starts by printing the exterior of the item, which is known as the casing, and then fills it according to the percentage of the infill. It holds to the previous layer for the cooling filament. After each sheet is finished, the surface is lowered to allow more space to be printed on a new substrate. The method will be repeated until the target is finished FDM printing permits of the fabrication of complicated geometry, such as hollow pills for gastric floating.

7.3.3 Pressure-Based Microsyringe (PBM)

PBM requires viscous or semisolid preparation as a starting material. The appropriate mixture of polymer, solvents, and other useful excipients can be suitable for printing. The material is put into a syringe, which has a piston or plunger at the top. A pressurized air piston is used to initiate the printing process (three to five bars). A 3D item is created by printing extruded strands onto a platform. Nozzle with a distinct diameter can be fixed to a syringe. Nozzle size depends upon the viscosity of the material. Viscosity needs to be analyzed to circumvent printing mistakes that can clog the nozzle. PBM does not require higher temperatures appropriate for thermosensitive drugs [10]. Post-processing drying of the object is required. PBM 3D printers additionally let in the manufacturing of "polypills" containing more than one active ingredients with different release profiles [11]. PBM-based fixed-dose formulations for HIV therapy are discussed by Siyawamwaya et al. [12]. PBM was utilized by Khaled et al. to make complicated multi-drug tablets with varied drug release patterns, as well as a polypill containing five separate active pharmaceutical ingredients with different release kinetics [13].

7.4 APPLICATIONS OF 3D PRINTING IN HEALTHCARE

7.4.1 Customization of 3D-Printed Dosage Form for Personalized Drug Dosing

3DP in pharmaceutical development has multiple possible avenues to introduce the evolution of medicinal drugs to undiscovered regions, creation of complex models of dosage forms, and customized medication [14]. 3DP benefits comprise a correct volume of the drug according to the patient's genetic profile and also have the ability to reproduce as well as the potential to manufacture medicine types with complicated drug release profile [15]. Tailored 3D-printed medicines would be beneficial for patients with multiple diseases with minimum side effects. Customization of a pill by incorporating more than one active agent mixture or multilayer tablet can be

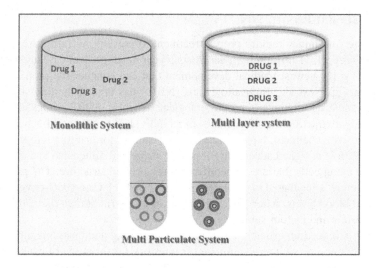

FIGURE 7.1 Personalized drug delivery according to patients' need.

fabricated through 3DP and simplifies the administration of dosage form (Figure 7.1). However, Liang et al. had used the FDM printer effectively for the development of mouthguards that can be tailored to the patient specification in the configuration and medication release rate [16]. 3DP fibers were biocompatible polylactic acid (PLA) and polyvinyl alcohol (PVA) mixtures filled with anti-inflammatory medication via HME (hot melt extrusion). A pharmacist would also get an opportunity to customize and print and dispense the dosage form to the patient.

7.4.2 ONCOLOGY APPLICATION

Cancer is the second most reason for mortality in the world. Sadly, new cases of infection may rise by 70% in the next 20 years [17]. 3DP is becoming a widely researched breakthrough in biomedical and pharmaceutical industries. It may be utilized for a variety of applications due to the decrease in manufacturing costs and time required for mass production, the accuracy of printing, and the ability to readily make minor alterations to items that would otherwise be impossible to do with other subtractive methods [18]. 3DP can contribute to lethal tumors such as glioblastoma by collecting cell samples from tumor-survivor patients and preparing bio-ink [19]. Bio-ink can be the basic material to create tumors on-chip and test for various medications and chemotherapies [20]. One of the fundamental targets in designing cancer models is the network of blood vessels which plays an important role in nutrients and oxygen. 3D bioprinting (3DBP) has a truly unique function that helps in operating tumor models and vasculature. Small channels can be formed inside the hydrogel mold by bio-ink followed by introducing endothelial cells to imitate vasculature inside the surface of microchannels [21]. Simultaneously, tumor microtissues can be embedded within the hydrogel matrix in the area of bio-printed small blood vessels permitting into tumor vascularization. It is a promising solution for cancer patients.

7.4.3 3D Bio-Printed Skin

Skin burns and injury require proper treatment to prevent mortality. According to the WHO report, 11 million burn accidents a year need medical care and treatment [22]. 3DP is the newest medical development that can conquer the problem. India-based start-up Next Big Innovation Labs (NBIL) has created human skin replica by understanding biological and physiological structure. 3DBP of the skin can be designed by extracting the skin cells from an individual patient and further the cells are mixed with chemical, bio-ink, and then transferred for printing into the 3D bio-printer. Skin is designed according to specific shape and dimension and structure is complete using cells that are embedded into the printed structure. The printed cellular structure is incubated to grow as human skin at $37 \pm 1°C$ and 95% oxygen similar to normal blood levels. After 14 days, cells grow and form human tissue by forming the epidermis and its four sublayers.

3DBP is indeed an efficient approach for the testing and customization of anti-tumor therapies, as well as a significant framework for the study of cancer biology and pathophysiology [23]. Bio-printed in vitro modeling of human skin will provide enormous monitoring opportunities for skincare and drug products and learn new physiological processes that are essential to human skin, potentially bridging the difference in traditional monolayers or 3DP between animal models. The cosmetic industry can test their products on bio-printed skin as an alternative to animal testing.

7.4.4 3D Printed Organs

Bioprinting technologies are increasing, and each big initiative is one step ahead to becoming a highly integrated and feasible solution. Research labs and testing facilities are bioprinting of human livers, kidneys, and hearts [23]. The challenge is to create them ready for transplantation and to have appropriate long-term alternatives. In particular, this approach may make it easier to cope with the scarcity of organ donations and to further try and understand those diseases.

3DP has a potential contribution to the precision of anatomical models to create significant features before operating patients. Furthermore, it leads to an effective procedure with minimum risk to the adjacent tissue or organ. An eight-month-old baby suffering from congenital heart disease was helped by 3DP in the preoperation strategy, which aided in creating a model of the heart for doctors to study and the surgery was successful. Tracheomalacia was observed in babies; it is a condition wherein it becomes difficult to breathe for a baby due to the abnormal growth of the trachea or windpipe. The splint was designed with the help of polycaprolactone [24]. Scientists in Newcastle University, UK, created 3DP synthetic cornea that can be replanted in patients. Scientists were able to create a 3D corneal model after examining and collecting data from the eye of volunteers. The suitable material is complicated to preserve the concave mold and stream the bio-ink into the mold. Bio-ink was made using alginate and collagen cells from healthy donors. Alginate was mixed with collagen and a special gel, and the mixture was formed as a unique gel that keeps the stem cells alive while creating cornea that is sufficiently rigid to retain structure and soft enough to be pulled off from a 3D printer [25]. Organ transplantation is a complicated as well as an expensive process as cells of a donor and receiver should match. The use of a cell from the patient's own body to create the substitute organ will possibly mitigate this concern.

7.4.5 Custom-Made Prostheses and Implants

Customized prostheses should match and adapt the wearer. Patients need to wait for weeks or months since manufacturing customized amputees by traditional method is time-consuming [26]. 3DP tremendously speeds up the process and also prints the prostheses inexpensively. The prosthesis that victims of natural catastrophes and war refugees were unable to afford due to cost has been resolved owing to 3DP [27]. To achieve normal and advanced, personalized prosthesis limbs and operating implants, 3DP is now used significantly in the healthcare sector that can print limbs often within 24 hours [28]. In orthopedics, where traditional implants are sometimes inadequate for certain patients, especially in complex situations, the ability to rapidly manufacture customized implants and prosthetic limbs solves the persistent issue. In the past, surgeons had to perform intricate graft material procedures with surgical tools and drilling to adjust implants to the correct form, scale, and fit by scraping bits of metal and plastic. Cranium has irregular forms; therefore, a cranial implant is difficult to standardize.

7.5 REGULATORY CONCERNS

The 3D implant is provided to the consumer by a healthcare professional. Implants are based on a range of variables, including measuring precision and surface finish. Implant performance depends on the capacity of the manufacturer (production of the implant), the 3D printer itself, and the 3D printer interface or program. In these circumstances, since the 3D printer producer does not communicate directly with the patient, the accuracy of the 3D printer is very critical in eliminating the impact of an error. Therefore, the 3D printer must be modified and regulated [29]. Every designed dosage form and the medical device is unique according to patients' needs. So, it is difficult for FDA to regulate every personalized device. The FDA might even approve 3D medical devices from abbreviated paths in addition to the conventional approval routes. These routes include paths for immediate use, pathways for humane use. For instance, an infant was suffering from tracheobronchomalacia in 2013. A biologically special tracheal splint was authorized by the emergency-use exemption pathway to save the child's life. However, it is unclear whether 3DP will move along these short lines of dosage. In August 2015, FDA approved the first ever 3D-printed tablet, Spritam® (levetiracetam), considering all the regulatory burdens linked with 3DP. Keeping in mind all processes and resources available at the moment, there seems to be no probability to include a standardized set of instructions to various types of printing processes [30]. The commodity is taken into consideration to approve new manufacturing processes. In more recent times, FDA drafted a guide for supporting technological requirements for 3D-printed devices that have the potential for 3D development [31].

7.6 ROBOTICS IN HEALTHCARE INDUSTRY

3DP coupled with robotics application reduces costs and speeds up turnaround time for the most laborious and repetitive operations.3DP is a new manufacturing method that is mostly performed by robots or in a robotic way. It's a fully automated procedure in which machines construct a thing layer by layer. At the same time, 3DP makes it possible to quickly make different types of robots and their parts. Currently, 3DP is done both for and by robotics.

Conventional manufacturing methods that required hours to perform a specific work are expected to be replaced by 3DP in the near future. There are several areas where 3DP may have a positive impact because of its rapid advancement.

7.6.1 WHEN IT COMES TO ROBOTICS AND AUTOMATION, HOW IS THE 3D PRINTING BEING USED IN THESE AREAS?

The use of 3DP and industrial robots is an effective way to meet the growing needs of manufacturers to make large objects on a substantial scale and with great precision, while also trying to make the production process more efficient.

As 3DP methods have become more advanced, it has become easier to combine them with industrial robots, which has led to the idea of "smart manufacturing."3DP is often used by manufacturers when they need to make a product that is accurate, fast, and in a small amount.

Industrial robots are benefiting from the introduction of 3D printers, which are helping to produce more efficient and effective robots [2]. With 3D printers, engineers can speed up the research and design phase, which means new, better robots can be made and used much faster than they would have with traditional prototyping methods.

The creation of 3D printers that use robotic arms was brought up during a discussion about the future of smart manufacturing. In and of itself, this is a significant new initiative in industrial robotics 3DP. In the 3DP sector, several companies are creating various types of robotic arms that are capable of carrying out 3DP operations.

7.6.2 IT'S POSSIBLE TO EASILY REPLACE A 3D-PRINTED JOINT USING A SURGICAL ROBOTIC SYSTEM

Monogram orthopedics has a new way to do joint replacements. This is how it works: The company's robotic arm precisely puts 3D-printed joint implants that are custom-made to match each patient's bone structure. Surgeons will be able to use less invasive procedures and better-fitting implants to replace joints, reducing infection risk, dislocation, and implant failure, according to the scientific consensus [32].

For monogram's custom-printed implants, the unique bone structure of each patient serves as a template for design. A CT scan of the joint to be replaced is the first step in the company's bespoke production process. With this data, an AI-powered preoperative interface helps the surgeon plan the surgery from every viewpoint using predictive algorithms. The implant is then 3D-printed using an automated manufacturing technology and shaped to fit flawlessly into the patient's joint.

Monogram's second component is a surgical robot arm. As the robot follows the surgeon's directions through a real-time virtual navigation display, human error is eliminated from the operation procedure. A tiny incision is made into the bone, and a milling drill is used to create a perfectly carved hollow. Using minimum or no cement, monogram claims that their implant is up to 270% more stable than a generic alternative.

7.6.3   3D Printing is Paving the Way for Robotic Surgery

Various phases of surgical operations have been improved by the use of additive manufacturing. Renowned physicians and institutions are increasingly preparing for and training using 3D-printed models. This was the case with Jadon and Anias McDonald's 27-hour separation, during which 3DP was employed for virtual planning during the pre-surgery procedure, as well as the production of skull caps for each baby.

In a number of pediatric head and neck surgical procedures, robotic surgery has been proved to be feasible and successful. On the other side, adoption has been limited. The development of new robotic platforms with application-specific tools, as well as improved preoperative surgical planning, may allow for a more seamless integration of robotic surgery into practice [33]. Computer-aided surgical planning techniques including 3D printing, virtual reality, a multi-objective cost function for approach optimization, mirror image overlay, and adaptable robotic instruments may bring greater value and applicability to present practice with additional study.

7.6.4   3D Printing and Patient-Specific Equipment in Orthopedic Robotic Surgery

Institutions are now producing patient-specific tools (robotic surgical instruments) that are tailored according to each patient's anatomy as technology and 3DP techniques progress [34]. Surgeons may utilize an MRI scanner to image a knee, for example, and use the resultant scan to designate where the robot can go and where it cannot. The robotic device has a set of goals and a specified route to follow.

7.6.5   3D-Printed Biological Soft Robots

Many different soft robots and actuators are being created to address contemporary biological problems. From surgical and diagnostic robots to active implants, robots may be used in a variety of ways. *In vitro* robots and *In vivo* robots are two of the primary kinds of biological soft robotics that may be categorized (operating inside a living organism). It is similar to the phrase "soft robot," which refers to at least one important soft component of the robot. To solely make soft robotic components, 3DP is often employed [35]. These parts are either impossible to make using traditional methods or very difficult to make using conventional methods.

7.6.6 In Vitro 3D-Printed Soft Robots

To function in a controlled environment outside the body of a human species, the term "in vitro" is used. The use of 3D-printed soft robots to solve biological challenges requiring operations carried out outside the body of animals is beginning to gain traction, because soft robotics can mold them to the intended body form for applications in surgery, exoskeletal active implants, therapy, diagnostics, artificial skin, and so on.

Using 3DP construction methods, similar in vitro soft robots have been built for active exoskeletal prostheses and rehabilitation. With a 3D-printed primary frame of the robot and the mood for soft actuators, a soft robot for gait rehabilitation of specialized rodents has been constructed [36]. A soft robotic-sensing unit containing printed soft sensors and electronics, as well as a 3D-printed mold assembly, has been designed for human gait assessment. To allow completely soft bio-robots, 3D-printed flexible electronics for monitoring vital signs have been designed. Active prostheses for amputees have been developed using 3DP-assisted fiber-reinforced soft actuators capable of following complicated trajectories. Australian researchers have developed 3D-printed flexure hinges for soft monolithic prosthetic fingers, which can be paired with soft actuators and a 3D-printed framework to create a fully functional prosthetic hand.

Also being researched are 3D-printed stretchy electronics, such as soft sensors and actuators, which may be paired with exoskeletal implants to perfectly imitate the bio-functionality of genuine human organs, enabling artificial robotic systems to detect things like touch and heat. Some researchers refer to the soft electronic sheet as "artificial skin" since it serves as a substitute for real skin on robotic organs.

7.6.7 In Vivo Soft Robotic

3D-printed soft robots are being utilized for a variety of medical procedures, including in vivo operations, organ implantation, targeted medication administration, and the diagnosis and treatment of a variety of diseases and ailments. Soft robotics has a wide range of in vivo applications since the internal body structure is too difficult to access with rigid materials [37]. In addition, most interior organs and tissues are sensitive and need similar repair and replacement structures. Tracheobronchial splints, 3D-printed soft and smart robotic implants, have been developed to treat tracheobronchial collapse in tracheobronchomalacia. The primary benefit of this active implant is its ability to be customized for each individual patient. 3D-printed soft micro-bio-bots can be used to monitor a wide range of conditions and diseases in the body. These robots can move through blood vessels and food tracks to get to parts of the body that aren't reachable by humans. They can offer vital information of the region through implanted sensors, as well as provide tailored drugs to the afflicted areas.

7.7 CONCLUSION

3DP is approaching growth. The development in the field of biomedical industry with the help of software applications will soon be modern new technology. The growing cognizance in 3DP will benefit and expedite growth. In order to allow continuous tracking, 3DP has begun to equip their machines with detectors and sensors to calculate dimensions. Over the last century, the primary focus has been on 3DP, a crucial new technology. 3DP has increasingly been embraced by companies such as healthcare, biomedical, aircraft, and vehicles. In upcoming days, it is predicted that 3DP can be merged with technology such as the Internet of Things, robotic sensors, machine learning (ML), and reduce the risk during the manufacturing process. A 3D smarter method would potentially produce a product with good quality and also support 3DP on the production line in proving performance. We briefly touched on the connection between robotics and 3DP throughout our chapter. Humans may now dive into the ocean and explore outer space thanks to the endless applications of robots in all aspects of our existence. Furthermore, advances in 3DP take robot production to the next level. We may envisage robots being built rapidly and cheaply using 3DP technology and its capabilities.

REFERENCES

[1] Awad, A., Fina, F., Goyanes, A., Gaisford, S., & Basit, A. (2021). Advances in powder bed fusion 3D printing in drug delivery and healthcare. *Advanced Drug Delivery Reviews, 174*, 406–424. doi: 10.1016/j.addr.2021.04.025

[2] Resources, A., & Insights, I. (2022). Building the Future with Robotic Additive Manufacturing. Retrieved 28 March 2022, from https://www.automate.org/industry-insights/building-the-future-with-robotic-additive-manufacturing

[3] Trenfield, S., Awad, A., Madla, C., Hatton, G., Firth, J., & Goyanes, A. et al. (2019). Shaping the future: Recent advances of 3D printing in drug delivery and healthcare. *Expert Opinion on Drug Delivery, 16*(10), 1081–1094. doi: 10.1080/17425247.2019. 1660318

[4] The Types of 3D Printing Technology | All3DP. (2021). Retrieved 21 January 2021, from https://all3dp.com/1/types-of-3d-printers-3d-printing-technology/

[5] Guide to Selective Laser Sintering (SLS) 3D Printing. (2021). Retrieved 21 January 2021, from https://formlabs.com/asia/blog/what-is-selective-laser-sintering/

[6] Fina, F., Goyanes, A., Gaisford, S., & Basit, A. (2017). Selective laser sintering (SLS) 3D printing of medicines. *International Journal of Pharmaceutics, 529*(1–2), 285–293. doi: 10.1016/j.ijpharm.2017.06.082

[7] Partee, B., Hollister, S., & Das, S. (2004). Selective laser sintering of polycaprolactone bone tissue engineering scaffolds. *MRS Proceedings, 845*. doi: 10.1557/proc-845-aa9.9

[8] Sadia, M., Alhnan, M., Ahmed, W., & Jackson, M. (2017). 3D printing of pharmaceuticals. *Micro and Nanomanufacturing2, 467*–498. doi: 10.1007/978-3-319-67132-1_16

[9] Alhnan, M., Okwuosa, T., Sadia, M., Wan, K., Ahmed, W., & Arafat, B. (2016). Emergence of 3D printed dosage forms: Opportunities and challenges. *Pharmaceutical Research, 33*(8), 1817–1832. doi: 10.1007/s11095-016-1933-1

[10] Azad, M. A., Olawuni, D., Kimbell, G., Badruddoza, A. Z. M., Hossain, M. S., & Sultana, T. (2020). Polymers for extrusion-based 3D printing of pharmaceuticals: A holistic materials–process perspective. *Pharmaceutics, 12*(2), 124. MDPI AG. Retrieved from http://dx.doi.org/10.3390/pharmaceutics12020124

[11] Curti, C., Kirby, D., & Russell, C. (2020). Current formulation approaches in design and development of solid oral dosage forms through three-dimensional printing. *Progress in Additive Manufacturing, 5*(2), 111–123. doi: 10.1007/s40964-020-00127-5

[12] Siyawamwaya, M., du Toit, L., Kumar, P., Choonara, Y., Kondiah, P., & Pillay, V. (2019). 3D printed, controlled release, tritherapeutic tablet matrix for advanced anti-HIV-1 drug delivery. *European Journal of Pharmaceutics and Biopharmaceutics, 138*, 99–110. doi: 10.1016/j.ejpb.2018.04.007

[13] Khaled, S., Burley, J., Alexander, M., Yang, J., & Roberts, C. (2015). 3D printing of tablets containing multiple drugs with defined release profiles. *International Journal of Pharmaceutics*, *494*(2), 643–650. doi: 10.1016/j.ijpharm.2015.07.067

[14] Ali, A., Ahmad, U., & Akhtar, J. (2020). 3D Printing in pharmaceutical sector: An overview. *Pharmaceutical Formulation Design—Recent Practices*. doi: 10.5772/intechopen.90738

[15] Prasad, L., & Smyth, H. (2015). 3D printing technologies for drug delivery: A review. *Drug Development and Industrial Pharmacy*, *42*(7), 1019–1031. doi: 10.3109/03639045.2015.1120743

[16] Al-Dulimi, Z., Wallis, M., Tan, D., Maniruzzaman, M., & Nokhodchi, A. (2020). 3D printing technology as innovative solutions for biomedical applications. *Drug Discovery Today*. doi: 10.1016/j.drudis.2020.11.013

[17] Worldwide Cancer Cases Expected to Soar by 70% Over Next 20 Years. (2021). Retrieved 20 April 2022, from https://www.theguardian.com/society/2014/feb/03/worldwide-cancer-cases-soar-next-20-years

[18] Charbe, N., McCarron, P. A., & Tambuwala, M. M. (2017). Three-dimensional bio-printing: A new frontier in oncology research. *World Journal of Clinical Oncology*, *8*(1), 21–36. https://doi.org/10.5306/wjco.v8.i1.21

[19] Glioblastoma News, Research. Retrieved 9 May 2022, from https://www.news-medical.net/?tag=/Glioblastoma

[20] A 3D Printer Loaded with Cancerous 'Ink' Churns Out Mini-Brain Tumours. (2019). From https://www.nature.com/articles/d41586-019-00907-2

[21] Kang, Y. (2020). 3D Bioprinting of Tumor Models for Cancer Research. From https://pubs.acs.org/doi/10.1021/acsabm.0c00791

[22] Burns. https://www.who.int/news-room/fact-sheets/detail/burns (accessed Apr 12, 2022).

[23] Caballar, R. (2021). How Bioprinting Skin Could End Animal Testing, Improve Skin Transplants. Redshift EN. Retrieved 22 March 2022, from https://redshift.autodesk.com/bioprinting-skin/

[24] 3D-Airway Printed Splint | Otolaryngology | Michigan Medicine. Otolaryngology. (2021). Retrieved 20 March 2022, from https://medicine.umich.edu/dept/otolaryngology/3d-airway-printed-splint

[25] First 3D Printed Human Corneas. Press Office. (2021). Retrieved 22 January 2021, from https://www.ncl.ac.uk/press/articles/archive/2018/05/first3dprintingofcorneas/

[26] The Successes and Failures of 3D Printed Prosthetics—PreScouter—Custom Intelligence from a Global Network of Experts. PreScouter. (2021). Retrieved 22 March 2022, from https://www.prescouter.com/2017/07/3d-printed-prosthetics/

[27] 3D Printing for Medical Use | MakerBot. MakerBot. Retrieved 23 March 2022, from https://www.makerbot.com/stories/medical/3d-printing-for-medical-use/

[28] How Far Have 3D-Printed Prosthetics Come and Gone in the Last Five Years? By AndrewWheeler. Engineering.com. (2021). Retrieved 23 March 2022, from https://www.engineering.com/story/how-far-have-3d-printed-prosthetics-come-and-gone-in-the-last-five-years

[29] Ventola, C. L. (2014). Medical applications for 3D printing: Current and projected uses. *P & T: a Peer-Reviewed Journal for Formulary Management*, *39*(10), 704–711.

[30] Fitzgerald, S. (2015). FDA approves first 3D-printed epilepsy drug experts assess the benefits and caveats. *Neurology Today*, *15*(18), 26–27. https://doi.org/10.1097/01.nt.0000472137.66046.b5

[31] Advanced Manufacturing. U.S. Food and Drug Administration. (2021). Retrieved 22 March 2022, from https://www.fda.gov/emergency-preparedness-and-response/mcm-issues/advanced-manufacturing

[32] This Surgical Robot Can Flawlessly Implant a 3D-Printed Joint Replacement. (2022). Retrieved 28 March 2022, from https://www.newsweek.com/sponsored/this-surgical-robot-can-flawlessly-implant-3d-printed-joint-replacement

[33] Ghazi, A., & Teplitz, B. (2020). Role of 3D printing in surgical education for robotic urology procedures. *Translational Andrology and Urology, 9*(2), 931–941. doi: 10.21037/tau.2020.01.03

[34] Medical 3D Printing Applications in Orthopedics. (2022). Retrieved 28 March 2022, from http://blog.peekmed.com/3d-printing-in-orthopedics

[35] Gul, J., Sajid, M., Rehman, M., Siddiqui, G., Shah, I., & Kim, K. et al. (2018). 3D printing for soft robotics—A review. *Science and Technology of Advanced Materials, 19*(1), 243–262. doi: 10.1080/14686996.2018.1431862

[36] Yap, Y., Sing, S., & Yeong, W. (2020). A review of 3D printing processes and materials for soft robotics. *Rapid Prototyping Journal, 26*(8), 1345–1361. doi: 10.1108/rpj-11-2019-0302

[37] Payne, C., Wamala, I., Bautista-Salinas, D., Saeed, M., Van Story, D., & Thalhofer, T. et al. (2017). Soft robotic ventricular assist device with septal bracing for therapy of heart failure. *Science Robotics, 2*(12). doi: 10.1126/scirobotics.aan6736

8 Self-Repeating Robotic Arm

An Experimental Learning

B. Nagamani, N. Subadra, Sathvik Parasa,
Hari Sarada, Ashrith Gadeela
Geethanjali College of Engineering and Technology,
Hyderabad, Telangana, India

CONTENTS

DOI: 10.1201/9781003121640-11

8.1   INTRODUCTION

A robotic arm, often known as an industrial robot, is frequently described as a 'mechanical' arm. It's a device containing a number of joints that move along an axis or spin in particular directions, comparable to a human arm. In reality, some robotic arms are anthropomorphic, meaning they strive to mimic the movements of human arms as closely as possible. They are usually programmed and designed to do specified tasks, most typically in manufacturing, fabrication, and industrial units. They can be little machines that conduct delicate, specific operations and can fit in one hand, or they might be massive enough to create entire structures.

Robotic arms were first developed to aid in mass production industries, most notably in the automobile industry. They were also adopted to reduce the danger of damage to employees and to complete repetitive jobs so that workers could focus on more sophisticated aspects of manufacturing. Earlier robotic arms were only used to perform simple, repetitive welding jobs. The function of robotic arms is evolving as technology advances, particularly robotic vision and sensor technology.

Today, technology is progressing in the same way as human desires are rapidly expanding. Every day, the work done to address these requirements makes life simpler. This project aims to develop a robot that is specifically designed for pick and place, which is a requirement in the business.

Many researchers contributed their research in this area. Few of them are D. L. Pieper (1), 1968, in his PhD thesis mentioned the kinematics of the robotic arm. H. Ueno and Y. Saito (2), 1996, explained the scheduling of human-type robot arm. Jayabala and Pradeep (3), 2016, developed a design of gesture-controlled robotic arm for industrial applications. Jennifer Bray and Charles F. Sturman (4), 2002, manufactured a wireless robotic arm with Bluetooth connection. L. Feng et al. (5), 1996, prepared a complete wireless system for mobile robots. M. Nakashima et al. (6), 1995, worked on human–robot interface and its applications. Mohammed Abu Qassem et al. (7), 2011, simulated a model of 5-DOF educational robot arm.

N. Sarkar et al. (8), 1997, developed a dynamic control in two arm manipulation. Nasr M. Ghaleb and Ayman A. (9), 2018, developed a model of 2-DOF robot arm. R. C. Luo and K. L. Su (10), 2003, studied and designed a multi-agent multisensor-based real-time sensory control system for intelligent security robot. Rahul Gautam (11), 2017, reviewed the development of industrial robotic arm. V. Potkonjak et al. (12), 2001, studied human-like behaviour of robot arms: general considerations and the handwriting task. Y. Huang et al. (13), 1997, developed a new type of machining robot with new type of driving mechanism. Yagna Jadeja and Bhavesh Pandya (14), 2019, developed a 5-DOF robotic arm.

Because automation in India is not inexpensive, no industrialist can afford to alter his unit manually or semiautomatically. The main goal of this chapter is to create a flexible and low-cost robotic arm that can perform pick-and-place operations. Micro servo motors and an Arduino Uno R3 microcontroller have been used to control the robot.

8.2 ROBOTICS AND AUTOMATION

8.2.1 WHAT IS ROBOTICS?

Any autonomously operated machine that takes the place of human effort, even if that doesn't seem like a human or performs activities in a humanoid manner is a robot. Robotics is a branch of mechatronics, electrical engineering, and computer science related to the design, construction, operation, and use of robots, as well as the controlling and processing of robots using computer systems. These mechanizations are associated with automated devices that can perform a variety of tasks, activities, situations, and processes in the absence of a person.

As technology progresses, the scope of what is referred to as robotics broadens. In 2005, three-quarters among all robots were employed in the automotive industry, constructing automobiles. These robots are generally made up of mechanical arms that are used to weld or screw on automobile parts. Robotics has evolved to include the design, production, and implementation of bots that can investigate Earth's harshest regions, robots that help public safety, and even bots that aid practically in every aspect of healthcare.

8.2.2 FUTURE OF ROBOTICS

Although robotic arms, systems, and equipment have been used in operating theatres since the early 1980s, their potential to aid surgeons with challenging tasks with higher precision is becoming more widely recognized.

Generations of people are growing more amenable to these sorts of advancements being employed for life-saving operations as technology continues to expand and evolve.

Consider your own arm while bearing in mind industrial robot arms. What is it capable of? It has the ability to bend. It can grip objects (with the help of the hand). It has the ability to lift objects. It has the ability to move objects. Industrial robot arms conduct processes that are similar to those performed by humans; however, they may be more efficient.

Industrial robot arms are used in a variety of sectors, including automotive, aerospace, electronics, food and beverage, and construction. These sectors have learned what a growing number of smaller, younger companies are discovering: employing industrial robotic arms for manufacturing boosts productivity.

8.2.3 AUTOMATION

The employment of electronics and computer-controlled devices to take control of operations is known as automation. The goal of automation is to increase efficiency and consistency. Automation, on the other hand, replaces labour in the vast majority of circumstances. Indeed, analysts now fear that new technologies will drive up unemployment rates dramatically in the future.

Robotic assembly lines are gradually taking over activities previously performed by humans in many industrial operations. Manufacturing is the process of transforming raw materials and components into completed things on a big scale, generally in a factory.

In substantially all sectors, automation involves several critical parts, systems, and job responsibilities. Manufacturing, transportation, facility operations, and utilities are all affected. National defence systems are also getting more and more automated.

Integration, installation, procurement, maintenance, and even marketing and sales are all areas where automation is used today in business.

8.3 STEPS INVOLVED IN CONSTRUCTION

8.3.1 Assembling the Circuit

8.3.1.1 Materials Required

8.3.1.1.1 Arduino

Arduino is a brand, a piece of hardware, a programming language, and a whole ecosystem of products all rolled into one. However, when we talk about Arduino, we're talking about open-source electronics prototyping platform in general.

An open-source electronics platform or board, as well as the software needed to programme it, is referred to as Arduino. Arduino is a platform for artists, designers, enthusiasts, and anybody interested in developing interactive objects or settings to make electronics more accessible. Because the hardware design is an open source, an Arduino board may be purchased, pre-assembled, or made by hand. Users can customize the boards to meet their own needs, as well as update and distribute their own versions.

The one used in this project is the Arduino Uno R3 board.

The Arduino Uno R3 is a microcontroller board that uses a detachable ATmega328 AVR microprocessor in a dual-inline-package (DIP) format. There are 20 digital input/output pins on it (of which 6 can be used as PWM outputs and 6 can be used as analogue inputs). It may be programmed using the Arduino computer software, which is simple to use. The Arduino has a large network of supporters, making it a simple way to get started with embedded electronics. The Arduino Uno R3 is the third and most recent iteration.

8.3.1.1.2 Servo motor

A servo motor is an electromechanical device that uses current and voltage to create torque and velocity. A servo motor is part of a closed loop system that provides torque and velocity as directed by a servo controller and is closed by a feedback device. The feedback device provides information to the servo controller, which modifies the motor action based on the requested parameters, such as current, velocity, or position.

There are many different types, styles, and sizes of servo motors.

The one used here is a dual axis servo motor.

8.3.1.1.3 Potentiometer

A potentiometer is a voltage divider that consists of a three-terminal resistor with a sliding or revolving contact. It operates as a variable resistor or rheostat when only two terminals, one end and the wiper, are employed.

Potentiometers, often known as pots, are electrical components that allow a human or a machine to adjust the voltage or resistance in one portion of a circuit directly. The voltage across the voltage sensor varies when the thumbwheel is spun. A resistive element is usually present in a potentiometer, and its resistance rises linearly with its length. The wiper pin glides over this resistive element, lowering resistance between one terminal and the wiper, while increasing resistance between the other terminal and the wiper. The wiper pin voltage is equal to the input voltage multiplied by the resistance of the wiper pin to ground divided by the overall resistance of the potentiometer.

8.3.1.1.4   LEDs

P-n junctions in light-emitting diodes (LED) are extensively doped. When forward biased, an LED will emit a coloured light at a certain spectral wavelength depending on the semiconductor material employed and the degree of doping. An LED is encased in a transparent cover, as seen in the illustration, to allow the emitted light to escape.

Compound semiconductor materials, which are made up of elements from groups III and V of the periodic table, are used to make LEDs (these are known as III-V materials). Gallium arsenide (GaAs) and gallium phosphide are two typical III-V compounds used to create LEDs (GaP).

Here, we have used LEDs for the sole purpose of the indication.

8.3.1.1.5   Batteries

An electrical battery is a power source that consists of one or more electrochemical cells connected to the outside world and may be used to power electrical devices.

When a battery is supplying electricity, the positive terminal acts as the cathode, and the negative terminal acts as the anode. The electrons will travel from the negative terminal to the positive terminal through an external electric circuit. When a battery is connected to an external electric load, it undergoes a redox reaction, which converts high-energy reactants to lower-energy products and provides the free-energy difference to the external circuit in the form of electrical energy. It also works as a mediator among the electric appliances and the current supply.

8.3.1.1.6   Resistors

A resistor is a passive electrical component with two terminals that employs electrical resistance to operate as a circuit element. Resistors are used in electronic circuits to control current flow, alter signal levels, divide voltages, bias active devices, and terminate transmission lines, among other things.

Resistors protect components from voltage spikes and guarantee that they receive the right voltage by producing a voltage dip. A certain voltage is required for each component in an electrical circuit, such as a light or a switch.

8.3.1.1.7   Connecting wires

A wire is a flexible metal strand that is generally cylindrical in shape. Electrical conductivity is established between two devices in an electrical circuit using wires. They have a very low resistance to electricity passing through them.

FIGURE 8.1   Bread board.

8.3.1.1.8   Bread board

In the world of DIY electronics, the breadboard (Figure 8.1) is the bread and butter. Even seasoned tinkerers utilize breadboards as beginning points for large-scale projects, as they allow newcomers to learn about circuits without having to solder.

You may have received a breadboard in your Arduino beginner kit or Raspberry Pi beginner kit if you are new to the world of DIY or microcontrollers. Let's take a look at what a breadboard is, where it originated from, and how you may utilize it.

8.3.1.2   Circuit Connections

The first step is assembling the circuit. There are four servo motors. A servo motor usually consists of three components which are signal (orange), power (red), and ground (black), the signal pin is connected to the digital pins 9, 10, 11, 12 of the Arduino.

The power and ground are given through an external source. Four batteries of 1.5 V each are used to connect the power and ground pins of the servo motor with the help of a breadboard.

These servo motors are moving components of the circuit, which moves in different angles set by the user. There are four potentiometers used for changing the angle of the servo motor.

A potentiometer has three pins: signal, ground, and power. The signal pins are connected to the Analog pins A0, A1, A2, A3 and the ground and power are connected to the GND and 5 V pins of the microcontroller. For switching on the circuit and to take the recordings for each action, push buttons are used. A push button has two terminals and one of the ends is connected to the digital output of six and seven and the other terminal is connected to the power supply of the Arduino with the help of a breadboard. Two resistors of 1k ohms are used to avoid an open circuit attached to the switches. For the detection of actions, four LEDs are used for recording four respective actions and another LED is used for indicating the ON.

8.3.1.3 Schematic Diagram

The circuit connections for the robotic arm's operation are shown in Figure 8.3.

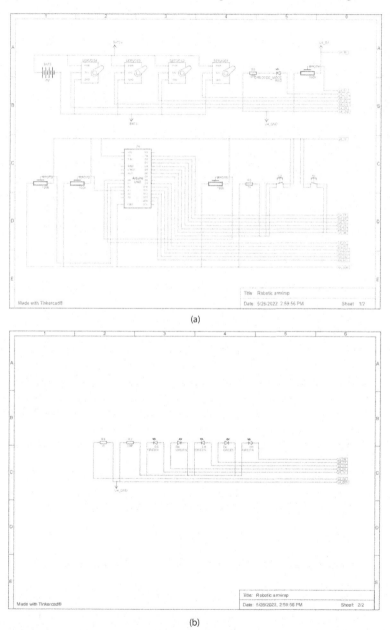

(a)

(b)

FIGURE 8.2   (a) Schematic diagram1 (b) Schematic diagram2

8.3.1.4   Circuit Diagram

The circuit connections for the robotic arm's operation are shown in Figure 8.3.

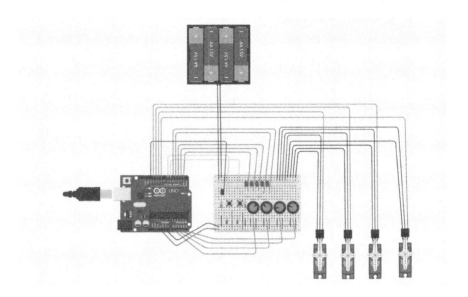

FIGURE 8.3   Circuit diagram.

8.3.2   PROGRAMMING THE ARDUINO

8.3.2.1   What Is C++

C++ is an object-oriented programming language with a wide range of applications. Bjarne Stroustrup developed it at Bell Labs in the 1980s. C++ is a language that is quite close to C (invented by Dennis Ritchie in the early 1970s). C++ is sufficiently similar to C that it will almost certainly build 99 percent of C programmes without modifying a single line of source code. C++, on the other hand, is a far more well-structured and safer language than C, since it is based on OOPs.

Some computer languages are created with a specific goal in mind. Java, for example, was created to operate toasters and other electronic devices. C was created to programme operating systems. Pascal was created with the intention of teaching correct programming skills. C++, on the other hand, is a general purpose programming language. The apt moniker "Swiss Pocket Knife of Languages" is well-deserved.

8.3.2.2   How Is It Used to Program Arduino in This Project?

C++ code is used to integrate hardware and software components. C++ code is a user-friendly programming language that allows the user to easily grasp and complete the programming within two functions; however, more functions can be utilized. The four servos are initialized at the top. The four servos, LEDs, and potentiometers are declared in the setup function by connecting to Arduino's analog or digital pins. Serial.begin is launching the serial monitor that shows digital output. The main line of code in the void loop function, where the subfunctions are run and the values are examined using if else conditions. The switch buttons are programmed in such a way that one switch is used to charge the Arduino board and the other to record events. Furthermore, the Readpos() method is used, in which the servo motor reads the value of the potentiometer specified by the user and sends it to the loop function. Finally, an automove() function is defined, which repeats recorded actions for an infinite number of times. This is how the C++ code is programmed to coordinate with the Arduino, which in turn coordinates with the hardware.

8.3.2.3 Code Used

```
#include <Servo.h>
int p[5][3];

Servo ser_0;
Servo ser_1;
Servo ser_2;
Servo ser_3;

int sn[10][5];
int num=0;

int c1,c2;

void setup()
{
  pinMode(A0, INPUT);
  pinMode(A1, INPUT);
  pinMode(A2, INPUT);
  pinMode(A3, INPUT);

  ser_0.attach(10);
  ser_1.attach(11);
  ser_2.attach(12);
  ser_3.attach(9);

  pinMode(13, OUTPUT);
  pinMode(1, OUTPUT);
  pinMode(2, OUTPUT);
  pinMode(3, OUTPUT);
  pinMode(4, OUTPUT);
  pinMode(5, OUTPUT);
```

```
  pinMode(6, INPUT);
  pinMode(7, INPUT);
  //Serial.begin(9600);
}

void loop()
{
    digitalWrite(13, LOW);
    readPos();
    delay(25);
    c1 = digitalRead(6);
    if(c1==1){
        c1=0;
        delay(500);
        while(1){

            digitalWrite(13, 1);
            readPos();

            c2 = digitalRead(7);
            if(c2==1){
                num++;
                for(int j=0;j<=3;j++){
                    sn[num][j]=p[j][0];
                }
                Serial.println(num);
                digitalWrite(num, HIGH);
                Serial.println(num);
                delay(200);
            }
            c1 = digitalRead(6);
            if(c1 == 1 || num>=5) {
                digitalWrite(13, 0);
                automove();
            }
        }
    }
}

void readPos(){
    p[0][0] = map(analogRead(A0), 0, 1023, 0, 179);
    p[1][0] = map(analogRead(A1), 0, 1023, 0, 179);
    p[2][0] = map(analogRead(A2), 0, 1023, 0, 179);
    p[3][0] = map(analogRead(A3), 0, 1023, 0, 179);
    if(p[0][1] != p[0][0]) {
      ser_0.write(p[0][0]);
      p[0][1] = p[0][0];
    }
```

```
      if(p[1][1] != p[1][0]) {
        ser_1.write(p[1][0]);
        p[1][1] = p[1][0];
      }
      if(p[2][1] != p[2][0]){
        ser_2.write(p[2][0]);
        p[2][1] = p[2][0];
      }
      if(p[3][1] != p[3][0]){
        ser_3.write(p[3][0]);
        p[3][1] = p[3][0];
      }
  }

void automove(){
    for(int j=1;j<=5;j++){
      digitalWrite(j,LOW);
    }

    while(1){
        for(int j=1;j<=5;j++){
            ser_0.write(sn[j][0]);
            ser_1.write(sn[j][1]);
            ser_2.write(sn[j][2]);
            ser_3.write(sn[j][3]);
            if( j<=5) digitalWrite(j, HIGH);
            delay(2000);
            if( j<=5) digitalWrite(j, LOW);
        }
    }
}
```

8.3.3 INTEGRATING THE HARDWARE

8.3.3.1 Physical Model

This Arduino-based robotic arm is meant to select and arrange things in response to human commands. It will choose and safely transport an object from the source to the destination. The arm's soft catching gripper will not exert any additional pressure on the objects. This nozzle has a payload capacity of 0.15 kg and can hold a box with dimensions of 50×50×70 mm.

Acrylic may be useful for low-stress structural sections such as robot bodies, depending on the weight of the robot and the needed shock and impact resistance.

Hence, we have used acrylic sheets for the body so as to make it lighter and easier on the objects it is holding.

8.3.3.2 Virtual Design

The virtual representation of the robotic arm is seen in Figure 8.4.

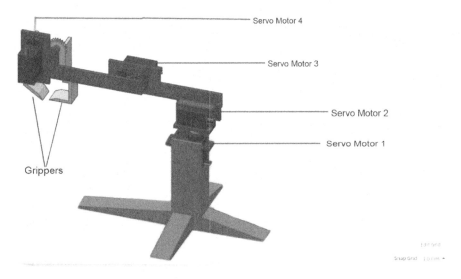

FIGURE 8.4   Virtual design.

8.3.3.3 Actual Design

The model's final result is shown in Figure 8.5. It is a test model.

FIGURE 8.5   Actual design.

8.4 WORKING OF THE ROBOTIC ARM

Push button 1 is pressed to configure the Arduino board. A red LED light begins to glow as soon as the push button is pressed. This shows that the board is prepared to carry out the next step.

The servo motor's hands may be modified with the aid of the potentiometer. Depending on the activities we want to take, the potentiometer value is increased and lowered for different angles. Actions are then stored by pressing push button 2.

A green LED light will immediately illuminate, indicating that the activity has been recorded. This step must be done for all servo motors in order to record appropriate actions for each servo motor.

Servo motors begin doing the repetitive moves as soon as four sets of actions are recorded. The recorded activities are repeated indefinitely in a loop. When push button 1 is hit again to end this, the arm stops moving and the previously recorded actions are deleted. The arm is now ready to record and execute new sequences of actions.

8.5 ADVANTAGES OF THE ARM

Advantages of a robotic arm are numerous, and they improve the efficiency and usability of a robotic arm; it can be used for multiple purposes. Precision and accuracy, as well as high performance, are two of the key advantages of using a robotic arm in the workplace. They perform exactly as they were programmed, with no deviations. The robotic arm has been designed to withstand weight and carry the desired payload with ease. It can be manufactured at a minimal price, allowing even small-scale enterprises to benefit from it. It may also be readily transported from one location to another, making it portable, as well as the controls are simple to operate.

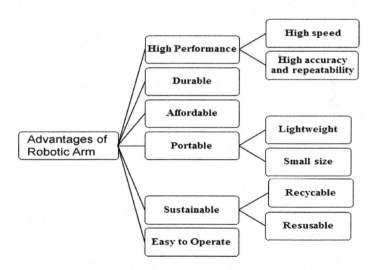

FIGURE 8.6 Advantages of the robotic arm.

8.6 FUTURE SCOPE

The development of robotic arms is extensive. In the not-too-distant future, arms will be able to execute any activity that humans do, and considerably better. The only limitation to its potential applications is one's imagination.

It can be a huge help to persons who are disabled or have lost their hands in an accident. The arm can be taught to respond to human commands and execute certain tasks. It is also feasible to create a precise gesture-controlled system.

Wearable gadgets can be used to convey commands and control arm motions. A person who has lost a hand in an accident can restart his or her life as before by using robotic arms.

As a result, robotic arms are adaptable and may be used in a variety of ways.

8.7 ROLE OF IoT

The Internet of Things (IoT) is an internetworking of physical devices, automobiles, organizations, and other items that are equipped with computers, software, sensors, actuators, and network connectivity to gather and share data.

The IoT plays an important role in robotics. It focuses on the communication, monitoring, and tracking of components. It is an interrelated network of all things that are connected to through the Internet, which include IoT-to-IoT device or IoT to any physical device, such as sensors or actuators. In case of robotic arm IoT provides an interface, which is used to control the robotic arm through a web page or an app. This can be done with the help of the Internet connected to the Wi-Fi router. The microcontroller receives the signal and the data is being processed and shown to the user.

8.8 CONCLUSION

This chapter describes our research and creation of a self-repeating robotic arm. It also describes the fundamental C++ code that is used to implement the arm. Even after many years of study, robotic arm applications remain limited to industries and are mostly employed in production units to increase productivity. These arms are incredibly smart and can do extremely accurate motions. Robotic arms have a wide range of general purpose and home uses that have yet to be fully explored. Cost is the fundamental barrier for large-scale robotic arms, and lowering it is a difficult challenge. However, our project is both cost effective and suitable for basic operations.

REFERENCES

1. D. L. Pieper. The kinematics of manipulators under computer control. PhD thesis, Stanford University, Department of Mechanical Engineering, 1968.
2. H. Ueno and Y. Saito, "Model-based vision and intelligent task scheduling for autonomous human-type robot arm," *Robotics and Autonomous Systems*, vol. 18, no. 1, pp. 195–206, 1996.
3. Pradeep Jayabala, "Design and implementation of gesture controlled robotic arm for industrial applications," *International Journal of Scientific Research*, vol. 3, pp. 202–209, 2016.

4. Jennifer Bray and Charles F. Sturman (2002). *Bluetooth: Connect Without Cables*. Upper Saddle River, NJ: Prentice-Hall.
5. L. Feng, J. Borenstein and D. Wehe, "A completely wireless development system for mobile robots," ISRAM Conference, Montpellier, France, May 27–30, 1996, pp. 571–576.
6. M. Nakashima, K. Yano, Y. Maruyama and H. Yakabe, "The hot linework robot system phase II and its human-robot interface MOS," Intelligent Robots and Systems 95. Human Robot Interaction and Cooperative Robots Proceedings. 1995 IEEE/RSJ International Conference, vol. 2, IEEE, 1995, pp. 116–123.
7. Mohammed Abu Qassem, Iyad Abuhadrous and Hatem Elaydi, "Modeling and simulation of 5 DOF educational robot arm," *International Journal of Mechanical and Mechatronics Engineering*, vol. 5, no. 3, 2011.
8. N. Sarkar, X. Yun and V. Kumar, "Dynamic control of 3-D rolling contacts in two-arm manipulation," *Robotics and Automation, IEEE Transactions*, vol. 13, no. 3, pp. 364–376, 1997.
9. Nasr M. Ghaleb and Ayman A. Aly, "Modeling and control of 2-DOF robot arm," *International Journal of Emerging Engineering Research and Technology*, vol. 6, no. 11, pp. 24–31, 2018.
10. R. C. Luo and K. L. Su, "A multi agent multi sensor based real-time sensory control system for intelligent security robot," IEEE International Conference on Robotics and Automation, vol. 2, pp. 2394–2399, 2003.
11. Rahul Gautam, Ankush Gedam, Ashish Zade and Ajay Mahawadiwar, "Review on development of industrial robotic arm," *International Research Journal of Engineering and Technology (IRJET)*, vol. 4, no. 3, pp. 1752–1755, 2017.
12. V. Potkonjak, S. Tzafestas, D. Kostic and G. Djordjevic, "Human-like behaviour of robot arms: General considerations and the hand writing task, Part I: mathematical description of human-like motion: distributed positioning and virtual fatigue." *Robotics and Computer-Integrated Manufacturing*, vol. 17, no. 4, pp. 305–315, 2001.
13. Y. Huang, L. Dong, X. Wang, F. Gao, Y. Liu, M. Minami, and T. Asakura, "Development of a new type of machining robot–A new type of driving mechanism," IEEE International Conference on Intelligent Processing Systems, 1997, vol. 2, IEEE, 1997, pp. 1256–1259.
14. Yagna Jadeja and Bhavesh Pandya, "Design and development of 5-DOF robotic arm manipulators," *International Journal of Scientific & Technology Research*, vol. 8, no. 11, November 2019.

9 Robotics Applications and Its Impact in the Global Agricultural Sectors for Advancing Automation

A. Kishore Kumar
Sri Ramakrishna Engineering College, Coimbatore,
Tamil Nadu, India

P.K. Poonguzhali, T. Nivethitha
Hindusthan College of Engineering and Technology,
Coimbatore, Tamil Nadu, India

CONTENTS

9.1 INTRODUCTION TO ROBOTICS IN AGRICULTURE

Presently, automation plays a significant role in a multitude of sectors to reduce human involvement, enhance productivity, and meet customer demand and contentment. Technology is being employed in a variety of domains, including automotive,

DOI: 10.1201/9781003121640-12

electronics, electrical, medical, defence, and agriculture, to increase efficiency and lowering human errors, costs, and downtime. Control and uncontrollable elements affect the majority of industries. However, because the majority of the process occurs within manufacturing plants, there are just some limited uncontrollable issues that determine the process. As in goods-producing sector, the output is primarily determined by a variety of factors that have direct or indirect influence on the output, such as the uncertainty of rainwater, storms, hot and cold waves, and so on. Traditional farming approaches are ineffective in these scenarios. Unconventional agriculture approaches combined with Internet of Things (IoT) technology have helped to lessen the effects of unpredictable factors and better comprehend meteorological conditions, which has resulted in higher agricultural output. There are no fore set boundaries for agriculture. The process of soil, planting, cultivating, watering, spraying, and harvesting is obviously included, but it is crucial to understand to what extent the process of soil, planting, cultivating, watering, spraying, and harvesting can be merged with certain methods of sorting and evaluating before they can be used in food manufacturing. Furthermore, animal-related tasks might extend beyond milking and shearing to shooting and processing, which is the first step in a long chain of operations that leads to the appearance of food products or manufactured apparel on supermarket shelves. With the adoption of the word 'robot' for the serial manipulator, robotics made its first significant debut in the manufacturing world. Manipulation and related kinematics were at the heart of the art, which turned into intelligent automation later. When the concepts of robotics are applied to agriculture, however, the emphasis must be on sensing than manipulation. When a tractor steers itself or a gate closes due to the detection of a feral pig, the answer to the question "Is this product innovation?" is not simple. Many developments in agricultural automation would be neglected if such applications were not included. Forestry is concerned with the collection of wood. Forestry machines are currently largely operated influenced by human pilots using global CAN-based intelligent devices, but in the future, these machines will become more autonomous and robot-like. A sensing system will be built into the machines, which will map the trees and locate the machine. As a result, information about the forest stand can be gathered, allowing for the operation of the motorized crane and loader, as well as manoeuvring and navigation. The majority of the forests in higher latitudes are natural and well-managed. There is efficient and sustainable silviculture, the science, art, and practise of caring for forests with reference to human objectives, particularly in Nordic countries. Before clear cutting, forest tree stands are thinned out and replacement trees are seeded or planted. In the future, autonomous machines for forestry will be a matter of study.

9.2 DISRUPTIVE TECHNOLOGIES IN AGRICULTURE SECTORS WITH THE COMBINATION OF ROBOTICS AND ARTIFICIAL INTELLIGENCE

Yet, the most significant impact on agriculture has been the increased knowledge of computational capabilities among farmers. During harvesting, cloud technology can record the yield and link it to a precise map of the property. Operations that were previously just automated can now be integrated and computerized, such as spray boom raising and rotating devices. The use of GPS (global positioning by satellite) for mapping and guide missions has exploded. Protocols used in generic radio communication techniques are similar to those used in smartphone applications or networked systems like Zigbee [1]. Automated systems enable for wireless connectivity of gates, cattle, or equipment, as well as automatic water trough replacement. In a forestry context, they can contribute equally to crew management and related conveyance. Certain radio technology, such as transponders, allows each creature to be tracked and traced, which helps with operations like milking and tracking "from the pasture to the plate." Hydraulic power has traditionally been used in agricultural equipment, but the inclusion of electronically controlled valves allows for autonomous steering and other "robot" functions. Transponders as well as other radio technology allow each species to be tracked and traced, which aids processes, such as milking and tracking "from the field to the tray." Agricultural infrastructure has historically employed hydraulic power, but the addition of digitally controlled valves permits for unsupervised driving and other "robot" activities.

Because of the possibility of legal penalties, it's unlikely that giant tractors will ever be free to wander unsupervised, but the day when collaborating crews of sensor-equipped "farmhand robots" will be used in the fields is getting closer [2].

9.3 ROBOTICS APPLICATIONS IN THE GLOBAL AGRICULTURAL SECTOR

Farmers are benefitting from the following robotics applications in the worldwide agriculture sector to increase revenue and productivity of various high-quality crops.

- Nursery planting: Nursery plants are marketed to consumers and gardeners, but for some crops, they are also the start of the food chain. Seeding, potting, and storing living plants in greenhouses are all automated by tech businesses.
- Crop seeding: Robotics and geo-mapping are combined in autonomous precision seeding. The soil parameters at each position in the field are depicted on a map. The seeds are then placed at precise positions and depths by the tractor's robotic seeding attachment, ensuring that each has the highest chance of germinating.
- Crop examining and investigation: Monitoring sensor and geo-mapping technologies facilitate agriculture to collect more data about the crops. Ground robots make it possible to acquire this data on their own.
- Fertilizing and irrigation: Robot-assisted precision irrigation helps save water by focusing irrigation on precise plants. Position robots find the way between the rank of crops in addition to water each plant openly from the root. Robots also have the advantage of being able to enter locations that other equipment is unable to.
- Crop weeding and spraying: Micro-spraying could cut the amount of pesticide used in crop production dramatically. Computer vision technology is used by micro-spraying robots to detect weeds and then spray a besieged plunge of insect repellent onto them.

Several technology companies have recognized the agriculture industry as a potential target for introducing farming-specific technologies to close the gap between farmer production and incomes. Robotic applications have been revolutionizing the agricultural sector in recent years and will continue to do so in the future as technology advances.

Agriculture is quickly sprouting into a modern engineering that attracts new qualified, production, and investors. Constantly developing awareness is enhancing both farmers' production capability including robotics and mechanization expertise. The assertion of considerably better assembling reveals the root of the issue. The world will require significantly more food, and farmers will be under enormous pressure to meet the need of scarcity of food. Farming bots assist farmers in a variety of ways to enhance output yields.

Drones, self-contained trucks, and robotic weaponry are all examples of how science is being applied in unique and novel ways.

Applications of agricultural bots: It's a massive task to keep track of vast agricultural areas. Farmers may now acquire a higher level of information concerning their yield than ever before thanks to new sensor and geo-mapping technologies. Drones and ground robots can acquire this information autonomously. One of the most important aspects of growing healthy crops is irrigating and fertilizing them. Clearly, this wastes water and reduces efficiency. Robot-assisted precision irrigation, for example, is able to assist in reducing wasted water by targeting certain vegetation. Position robots travel through rows of fields on their own, pouring water directly at the roots of every plant.

Robots also have the ability to reach locations where other devices cannot. Robotic innovations will also aid with the agriculture industry's personnel need. Farmers can focus more on boosting overall crop yields by using farming robots to mechanize sluggish, recurring, and boring jobs.

- Picking and harvesting
- Controlling weed
- Self-directed cut down, trimming seed, spraying, and thinning
- Phenotyping
- Packing and packing
- Effectiveness platforms are some of the most prevalent uses for robots in agriculture.

As of the exactness and swiftness with which robots can boost capitulate and diminish waste from yield left in the meadow, bring in and picking is one of the in-style robotic applications in cultivation. Automating these applications, on the other hand, can be difficult. Farmers are increasingly using harvesting and picking robots, but there are dozens of other creative ways the agricultural business is using robotic automation to boost output yields. Food demand is outstripping available cropland, and it is up to farmers to bridge the gap. Agricultural robots are assisting them in this endeavour.

On the surface, crop harvesting appears to be ripe for automation. Physically demanding and extremely repetitive labour is the most effectively targeted by the robot revolution.

To achieve operative request as well as protection in self-sufficient agricultural systems, various operations are necessary to be performed instantaneously, counting the system, the crop pitch, and exterior factors, such as human supervisors. The specific activities to make up a completely autonomous agricultural system include absolute or relative field localization, hindrance and stimulating section finding, communication with exterior operators or other self-directed units, sovereign navigation or remote operation, and site-specific applications.

Large-scale agriculture and dairy operations benefit from collaborative and cooperative behaviour between robots because numerous activities can be completed simultaneously, resulting in superior economies of scale. In a systematic approach, dissimilar category of robots and self-directed systems can now be fetched together. Because of their limited payload, operational time, and endurance, UAVs (Un-Manned Air Vehicle) or drones are ideal for aerial surveillance but not for ground tasks like spraying. As a result, ground and airborne vehicles must work together to complete their missions. Large-scale farm automation will considerably benefit from the control of several robots via a centralized software platform.

Companies in agricultural robot market are likely to employ multimodal robot systems extensively during the forecast period, owing to their capacity to run swarms of robots in farms to conduct numerous agricultural applications, such as spraying and weeding, simultaneously. Investing in multimodal systems is projected to increase revenue streams for agricultural robot manufacturers.

9.4 DEPLOYMENT OF AUTOMATION SYSTEMS IN GLOBAL AGRICULTURAL SECTORS-TECHNOLOGICAL FOCUS

The agri-robotics clutch has recently engrossed on categorizing applications where automating repetitive procedures is more well organized or active than using anthropoid or hefty appliance. Robotic plinths operating on the ground near the field or at advancement and enhanced management, particularly collaborating or tangible and scapes, are needed, for example, to harvest soft fruit. Human operators can have an eye in the firmament for comment and operation planning by using diverse multimodal systems for synthesis ground-centred and airborne automobiles. Large scale of land and fruit yields profit from cooperative and supportive comportment because responsibilities can be completed in parallel, resulting in cost savings. Due to challenges in fertilization, water management, and obtaining information on soil carbon content in the UK environment, upgraded certain management techniques retaining secluded stages, automation will become essential.

Automation is already used in a diversity of related fields in significant parts of industrialized manufacture of food in the UK [3]; activity of research is required to be more convincingly incorporated into agriculture. Multifaceted nutriment production and software systems are used to manage the food chain, which rely on reliable data concerning the position, eminence, and the amount of agronomic products. Earlier, robotics and mechanization cast-off in the food dispensation business. The practise of collecting large amounts of data and combining it with remote sensing, such as in the monitoring of raw resources for industrial manufacture, is gaining traction.

Elevating the cumulative and eminence of yields produced are feasible to convert the agricultural division in the United Kingdom. Technologies from connected fields, such as the IoT, Big Data, and artificial intelligence can be combined with sovereign system skills to repeatedly fuse and deduce composed data, evaluate yield progress and appropriate interventions in reaction to unexpected actions and changes in yield environments is maintained.

9.5   MACHINERY VISUALIZATION

A new generation of clever, versatile, durable, compliant, interrelated autonomous solutions that operate in harmony with human coworkers is essential for our long-term technological strategy in agricultural and plant products [4]. Teams of multifunctional robotic structures can orient that organizes its motions in tandem by and inside current agricultural and food systems. Digital farmland and commercial robots with interchangeable devices, such as limited solutions, breakthrough soft robotic grasping technology, and sensors, would enable agriculture to sustainably flourish and boost manufacturing efficiency along the food chain. Future agri-robotic systems will incorporate cognitive computing approaches to increase their production.

In the medium term, research into alternative frameworks for food production, including improvements from areas such as upright cultivating, will help to address the economic escalation of agribusiness while also ensuring the environment, food quality, and human well-being. Encouraging the move to robotization, whereas full mechanization is frequently hailed as the extreme point in innovative advancement, and long-term agribusiness frameworks may see exceptionally distinctive from those of nowadays, as it were exceptionally few expansive companies can bear the disturbance of full computerization. Therefore, to accomplish this long-term vision will necessitate a progressive move from the current cultivating hones, and most ranchers will require innovations than can be presented step by step, nearby and inside their existing frameworks. Besides, whereas a few developing automated advances are as of now accomplishing or illustration nearer the strength and cost-effectiveness required for real-world sending, other advances are not however at that arrange. For illustration, delicate natural product picking still necessitates essential investigate in detecting, control and soft robotics [5]. Hence, at slightest within the brief agrarian automatic technology term, the collaboration of people and robots is principal to expanded efficiency and nourishment quality. There are a number of moderately low-cost stages reachable presently that are certified for utilize nearby human specialists.

9.6   APPROACHING ROBOTIC AGRICULTURE SYSTEMS: ENABLING TECHNOLOGIES

A wide run of innovations will empower the transition of rural mechanical technology into the field. A few technologies will be created specifically for agriculture, while other innovations created for other areas may be adjusted to the rural space for the time being, for example, independent vehicles, manufactured insights, and machine vision. Here, we briefly survey the current status, opportunities, and benefits of different empowering technologies from equipment to program, multi-robot frameworks, and human robot frameworks. Figures 9.1 and 9.2 showcase the

FIGURE 9.1 Moisture seeding planting irrigation gaming agricultural Robot – side view. (Dr. A. Kishore Kumar [March 2019]. "MSPIG AGRI BOT [Moisture Seeding Planting Irrigation Gaming Agricultural Robot]," *EPRA International Journal of Multidisciplinary Research [IJMR] Peer Reviewed Journal*, vol. 5, no. 3. SJIF Impact Factor: 5.148 ISSN [Online]: 2455–3662.)

FIGURE 9.2 Moisture seeding planting irrigation gaming agricultural Robot – top view. (Dr. A. Kishore Kumar [March 2019]. "MSPIG AGRI BOT [Moisture Seeding Planting Irrigation Gaming Agricultural Robot]," *EPRA International Journal of Multidisciplinary Research [IJMR] Peer Reviewed Journal*, vol. 5, no. 3. SJIF Impact Factor: 5.148 ISSN [Online]: 2455–3662.)

Moisture Seeding Planting Irrigation Gaming Agricultural Robot from the side view and top view, respectively.

9.7 ROBOTIC PLATFORMS

Agrarian stage will be separated by space task, precise robots outlined to achieve particular assignment for given trim-stipulated space, bland stages intended to accomplish a few errands totally diverse domain likely to show critical parts. Ranches have a wide range of foundations; basic robots may be able to work in a certain area and to a limited extent across different ranches. The majority of existing automated stages are not resistant to everyday conditions, which is a corporate task. The majority of present exploiters are less concerned with conservatory moisture. Mechanics and microchip technology are two terms that are often used interchangeably. The development of quick testing methodologies and minimal processors has sparked a surge of additive manufacturing processes for "maker" machinery, enhancing possibility of moderate automated phases for a diverse submission. Embedded programme allows for highly customizable in implementation stages, be adjusted to a variety of roles and can use common hardware modules. While similar methods are used in UAVs and relatively small robots, lots of room for mechanical technology can be used in agro-food on a much larger scale. Control maintenance levels would be capable of functioning on all days; vigour and reliability issues will be addressed as prototypes progress to robust commercial stages.

Mobility bots must be capable of moving in dynamic, semi-structured environments. Ethereal vehicles must perform for lengthy periods in a variety of meteorological conditions, whereas terrestrial bots must travel on lumpy, homogeneous, and muddy terrain. Today's agricultural robots are primarily created by imitating business models and enhancing existing technologies, such as independent tractors. Therefore, they must not continue to be fully progressed in their activities, or they run the risk of being restricted to current systems. Surface segments must be able to play an important role on rails and flooring in hatcheries, on crushed rock or grass in underground burrows, and in incredibly sloppy and complicated land in open areas, multiple rotors, or a fixed wing frameworks are required for UAVs to fly.

Robotic arms will be required for the following errands in upcoming agriculture, including swapping skilled human worker, lowering prices for improving performance, and completing processes more carefully than present larger machinery such as slaughter extractors [6]. This course requires ongoing effort, by easy grippers being cast-off for test effort on tomatoes, raspberries, strawberries, mushrooms, and sweet peppers. Additional uses, like broccoli reaping, carried out with spiteful devices will need to be handled gently and the chosen edit's output will be limited. There are supplementary duties to collection in the meadow, and for guaranteed crops, where controllers might play a significant role. Motorized harvesting, precise scattering, types of inspection, and treatment are included. Increasing productivity in food control claims for huge mechanized stockrooms will need the use of swindlers.

9.8   SENSING AND PERCEPTION

The incorporation of products that are innovative into robot navigation frameworks opens up the possibility of new metrics that would otherwise be absurd. For example, present research uses infinite beam sensors adjusted from the dormant COSMOS techniques to map expansive zone fields for bulk dampness by various robots [7]. Notable advances in inaccessible detecting skills based on lackeys or drones give doors to monitoring trim development state with spectacular time and space resolutions at an affordable cost. Cattle farmers can access a variety of free datasets [8]. Robotic technologies are advancing the potential of legal soil testing with geotagging and speedy outcomes from sampling sensors, as well as protected test collection for further research in a systematic and uncontaminated manner. Lightweight bots and on-board secure collecting devices will further improve the regulatory efficiency and reliability of robot-assisted terrestrial organization systems.

9.9   LOCALIZATION AND MAPPING

Routing and placement related to transmission of real-time kinematics consenting correctness of centimetres related to computerized placing of huge agricultural technology, such as human-operated vehicles and combination gatherers, the use of GPS routes in farming has become nearly ubiquitous. Furthermore, methods relying solely on information management of the GPS signal have appeared to provide the same precision without the use of supplementary radio signals. Unmanned vehicles do not use GPS to keep precise location data because precise localization frameworks can be used with visible fiducial markers, auditory, or wireless signals, dependent on the quickness and correctness necessary. To ensure the harmless process of automated cars, sensor data is also essential in recognizing objects and threats in the field.

For example, driving autonomous vehicles to follow trim marks in millimetres or to follow tracks cleaned away by previous tractor operations is appealing. Multimodal frameworks that combine GPS, INS, LiDAR, vision, and other sensors have the ability to provide precise and robust configurations without the use of in-field frameworks like beacons. Several attempts to employ wildflower mapping principles by reflexively capturing its geographical position. Agriculture bots could also remain outfitted through layout of different classifiers, which use computer vision to anticipate the width and species of various weeds. Other approaches rely on a detailed semantic classification of weed images collected by drones [9].

9.10   CROP MONITORING

Using both ground and etheric stages, the third assessment can be precisely incorporated into agronomic practices using information combination systems. To provide surveillance and engrossment outputs at the specific plant dimension, it is integrated with virtual reality or augmented reality frameworks. Enduring data gathering improves crop forecasting concluded time, for instance, the evolution of edit canopy is traced and allowed for a more accurate projection of future growing conditions. With detachable (fawning) or partial remote detecting developments, both terrestrial and aerial structure stages open up new possibilities for permitting restricted, high signal to noise, high determination detecting. These robot stages at their best have the ability to extract reflectance and transmission from near vicinity (within tens of millimetres). Vision-based tasks for edit checking include phenotyping [10], determining when different vegetation are ready for harvest, and quality evaluation, which includes spotting the onset of illnesses.

Hyperspectral imaging data helps to correct inaccurate measurements caused by surface topography and the incorporation of human-altered tissues. At a further level, using computerized controllers to deploy sensors around crops or animals might allow reactions to be tested and inspected using artificial boosts. For example, by aiming a beam of light at specific ranges and changing the range centred, it is possible to move snaps that empathize into precise fragments like stems, early shoots, elderly shoots, and so on, which may then be spotted using hyperspectral imaging which allows for the recovery of substantially more prominent phenotypic data from over plants than is possible with static fixed imaging detector alone. Furthermore, the cell architectures and pathways of action within organic foods, vegetables, and meats can be examined non-destructively in order to make high-level decisions, such as mapping superficial bruises [11–12].

9.11 ROBOTIC VISION

Artificial learning approaches have a lot of potential for improving food production's environmental structure independence. To enable scene analysis and safe operation of automated frameworks in the field, visualization systems are also required for location, dissection, organization, and able to follow substances, such as organic foods, plants, cattle, individuals, and so on, as well as semantic division of crops versus weeds [13–14]. Approaches based on 3D point cloud analysis, such as those resultant from stereo symbolism, provide a notable assurance for achieving robust perception in difficult rural environments [15]. In pigs, cattle, and poultry, machine vision is having an influence in domestic nursing, such as for weight estimation, health status monitoring, and disease location [16–17]. Precision agriculture is a crucial way to achieve greater yields by utilizing the natural deposits in a diverse environment [18].

Deep learning from real-world datasets is often used in automated vision, with techniques like deep neural systems [19–21] gaining traction and increasing the likelihood that robots would share their data by benefitting from huge data. Flexible learning, encouraging adjustment to periodic changes, modern emerging illnesses, and pest perception in hard rural conditions are exposed problems in mechanical vision and machine discerning for automated farming. In pigs, cattle, and poultry, automation at present is having a control in animal nursing, such as heaviness approximation, physique condition monitoring syndrome detection.

9.12 CONCLUSION

Basic agricultural procedures, such as increasing yield performance and minimizing the use of potentially dangerous pesticides, have begun to be improved by robotics and other innovations. Modern agriculture robot systems are also being created to enable the integration of many technologies while providing modularity, flexibility, and adaptability. Integrating the latest technology in agricultural autonomous mobile robots, such as high-definition cameras and laser systems, allows the information to be blended to improve the sensory system's performance in terms of improved accuracy, resilience, and complementary data. According to World Bank data, agricultural employment has decreased by 15% globally in the last decade.

With an ageing farmer population significantly limiting the supply of physical labour, the labour shortage has become a global issue. Younger generations are less inclined to pursue farming, and children from farming households frequently relocate to urban areas in search of better job opportunities. In today's agricultural sectors, robotic automation will play a significant role. Agriculture's collaboration with technology could be the most transformational of all time. Agriculture is a sector that contributes significantly to a country's economic prosperity and stability. The transformation of agriculture from a low-tech employment to a full-fledged high-tech enterprise has shifted development criteria, and standards are continuing to rise. In the agricultural sector, robotic automation will allow farmers to focus more on the entrepreneurial parts of their business rather than the manual labour needed. These will result in labour and time savings as well as an increase in agricultural yield, demonstrating that technological advancement is omnipresent in our lives. Farmers will be able to spend less time in the business and more time on the business as a result of it.

REFERENCES

[1] D. Gislason and T Gillman, "ZigBee wireless sensor networks," *Dr. Dobb's Journal*, 2004, vol. 29, pp.40–42.

[2] J. Billingsley, "Low cost GPS for the autonomous robot farmhand," in *Mechatronics and Machine Vision: Research Studies Press*, 2000, pp. 119–125.

[3] A. Halme, K. Hartikainen, and K. Kärkkäinen, "Terrain adaptive motion and free gait of a six-legged walking machine," *Control Engineering Practice*, 1994, vol. 2, pp. 273–279.

[4] A. Bechar and C. Vigneault, "Agricultural robots for field operations: Concepts and components," *Biosystems Engineering*, vol. 149, pp. 94–111, Sep 2016. https://doi.org/10.1016/j.biosystemseng.2016.06.014

[5] A. Bechar and C. Vigneault, "Agricultural robots for field operation. Part 2: Operation and systems," *Biosystems Engineering*, vol. 153, pp. 110–128. https://doi.org/10.1016/j.biosystemseng.2016.11.004

[6] L. Grimstad and P. From, "The Thorvald II agricultural robotic system," *Robotics*, vol. 6, no. 4, p. 24, Sep 2017. https://doi.org/10.3390/robotics6040024

[7] STFC Research Grant ST/N006836/1. (2016). Synthesis of remote sensing and novel ground truth sensors to develop high resolution soil moisture forecasts in China and the UK. http://gtr.ukri.org/projects?ref=ST%2FN006836%2F1

[8] European Space Agency. (2018). Sentinel online. https://sentinel.esa.int/web/sentinel/home

[9] C. Zhang and J. M. Kovacs, "The application of small unmanned aerial systems for precision agriculture: A review," *Precision Agriculture*, vol. 13, no. 6, pp. 693–712, July 2012. https://doi.org/10.1007/s11119-012-9274-5

[10] M. P. Pound, J. A. Atkinson, A. J. Townsend, M. H. Wilson, M. Griffiths, A. S. Jackson, A. Bulat, G. Tzimiropoulos, D. M. Wells, E. H. Murchie, T. P. Pridmore, and A. P. French, "Deep machine learning provides state-of-the-art performance in image-based plant phenotyping," *GigaScience*, vol. 6, no. 10, pp. 1–10, Aug 2017. https://doi.org/10.1093/gigascience/gix083

[11] K. Kusumam, T. Krajník, S. Pearson, T. Duckett, and G. Cielniak, "3D-vision based detection, localization, and sizing of broccoli heads in the field," *Journal of Field Robotics*, vol. 34, no. 8, pp. 1505–1518, Jun 2017. https://doi.org/10.1002/rob.21726.

[12] M. Barnes, T. Duckett, G. Cielniak, G. Stroud, and G. Harper, "Visual detection of blemishes in potatoes using minimalist boosted classifiers," *Journal of Food Engineering*, vol. 98, no. 3, pp. 339–346, Jun 2010. https://doi.org/10.1016/j.jfoodeng.2010.01.010

[13] S. Haug, A. Michaels, P. Biber, and J. Ostermann, "Plant classification system for crop/weed discrimination without segmentation," in IEEE Winter Conference on Applications of Computer Vision. IEEE, Mar 2014. https://doi.org/10.1109/wacv.2014. 6835733

[14] P. Lottes, M. Hörferlin, S. Sander, and C. Stachniss, "Effective vision-based classification for separating sugar beets and weeds for precision farming," *Journal of Field Robotics*, vol. 34, no. 6, pp. 1160–1178, Sep 2016. https://doi.org/10.1002/rob.21675

[15] P. Bosilj, T. Duckett, and G. Cielniak, "Connected attribute morphology for unified vegetation segmentation and classification in precision agriculture," *Computers in Industry, Special Issue on Machine Vision for Outdoor Environments*, vol. 98, pp. 226–240, 2018.

[16] L. N. Smith, W. Zhang, M. F. Hansen, I. J. Hales, and M. L. Smith, "Innovative 3D and 2D machine vision methods for analysis of plants and crops in the field," *Computers in Industry*, vol. 97, pp. 122–131, May 2018. https://doi.org/10.1016/j.compind.2018.02.002

[17] M. Hansen, M. Smith, L. Smith, K. A. Jabbar, and D. Forbes. (June 2018). "Automated monitoring of dairy cow body condition, mobility and weight using a single 3D video capture device," *Computers in Industry*, vol. 98, pp. 14–22, Jun 2018. https://doi. org/10.1016/j.compind.2018.02.011

[18] K. A. Jabbar, M. F. Hansen, M. L. Smith, and L. N. Smith. (January 2017). "Early and non-intrusive lameness detection in dairy cows using 3-dimensional video," *Biosystems Engineering*, vol. 153, pp. 63–69, Jan 2017. https://doi.org/10.1016/j.biosystemseng. 2016.09.017

[19] M. F. Hansen, M. L. Smith, L. N. Smith, M. G. Salter, E. M. Baxter, M. Farish, and B. Grieve, "Towards on-farm pig face recognition using convolutional neural networks," *Computers in Industry*, vol. 98, pp. 145–152, Jun 2018. https://doi.org/10.1016/ j.compind.2018.02.016

[20] I. Sa, Z. Ge, F. Dayoub, B. Upcroft, T. Perez, and C. McCool, "DeepFruits: A fruit detection system using deep neural networks," *Sensors*, vol. 16, no. 8, p. 1222, Aug 2016. https://doi.org/10.3390/s16081222

[21] A. Kamilaris and F. X. Prenafeta-Boldú, "Deep learning in agriculture: A survey," *Computers and Electronics in Agriculture*, vol. 147, pp. 70–90, Apr 2018. https://doi. org/10.1016/j.compag.2018.02.016

Index

Printed in the United States
by Baker & Taylor Publisher Services